TROUT FLY RECOGNITION

Photograph W. J. Howes

THE AUTHOR AT HIS FLY-TYING BENCH

Trout Fly Recognition

JOHN GODDARD

WITH DRAWINGS BY CLIFF HENRY
AND A LIST OF OVER 150 ARTIFICIAL FLY DRESSINGS
BY JOHN VENIARD

ADAM & CHARLES BLACK · LONDON

FIRST PUBLISHED 1966
SECOND EDITION 1971
BY A. AND C. BLACK LIMITED
4, 5 AND 6 SOHO SQUARE LONDON W.I

© 1966 JOHN GODDARD

ISBN 0 7136 1288 6

DEDICATION

TO MY WIFE EILEEN

for her understanding and assistance

*Reproduced and Printed in Great Britain by
Redwood Press Limited, Trowbridge & London*

CONTENTS

	PAGE
Introduction by David Jacques	9
Author's Preface	11
Acknowledgements	13

CHAPTER

I.	The Four Major Groups	15
II.	The Upwinged Flies (Ephemeroptera)	24
	Appendix to Chapter II. The Baëtidae	31
III.	Two simple Identification Keys for the Duns and Spinners of the Ephemeroptera Order (The Upwinged Flies)	35
IV.	Detailed Descriptions of the Upwinged Flies (Ephemeroptera)	50
V.	Fishing Information on the Upwinged Flies and their distribution	82
VI.	The Flat-Winged Flies (Diptera)	105
VII.	The Sedge-Flies (Trichoptera)	109
VIII.	The Stoneflies (Plecoptera)	120
IX.	Sundry Insects and Other Fauna	126
X.	Nymphs of Ephemeroptera and Plecoptera	132
XI.	Collecting and Photography	148
XII.	Artificials suggested to match the Natural	157
	Appendix "A"—Artificial Flies by John Veniard	162
	Appendix "B"—Special patterns by the Author	191
	Appendix "C"—Body Parts of Flies	193
	References	196
	Index	197

ILLUSTRATIONS

In colour

PLATE		FACING PAGE
I.	Duns, etc. ×1·6	50
II.	Duns and Spinners, etc. ×1·6	60
III.	Duns and Sedge-Flies. ×1·6	70
IV.	Duns and Sedge-Flies. ×various	76
V.	Sedge-flies and other Fauna. ×1	86
VI.	Spinners. ×1	96
VII.	Spinners. ×1	104
VIII.	Spinners and Stoneflies, etc. ×various	124
IX.	Nymphs and Stoneflies. ×various	132
X.	A Trout's Eye View (Photographs of Naturals and matching artificials from underwater. ×1)	162

		BETWEEN PAGES
XI	Special Patterns	
XII	Special Patterns	192-3
XIII	Special Patterns	

Photographs and Drawings referred to in the Text

Fig. 1.	The Four Major Groups	15
Figs. 2–5.	Showing difference between Male and Female Upwing fly	26
Figs. 6–10.	Size Key for Upwinged Flies	36
Figs. 11 and 12.	Forewing and Hindwings	38
Figs. 13 and 14.	Venation and Hindwings	40
Fig. 15.	Ventral markings of *Ephemerella notata* and *Ecdyonurus insignis*	80

Figs. 16–31.	Venation of Hindwings of Upwinged Flies	80
Figs. 32–45.	Venation of Forewings of Upwinged Flies	between pp. 80 and 81
Fig. 46.	Ventral markings of Mayflies	84
Figs. 47–49.	Leg Spurs and Wing of Black Gnats	106
Fig. 50.	Lightweight Collecting Net	149
Fig. 51.	Entomological Names for various parts of Fly (Appendix)	194
Fig. 52.	A Mayfly—showing fisherman's names for various parts (Appendix)	194

Other Illustrations

The Author	*frontispiece*
Caddis Cases and Caënis Transposing	*facing p.* 112
The Camera	*facing p.* 128
The Upper Kennet and Photographic Equipment	*facing p.* 129

INTRODUCTION

Very few fly-fishermen have not heard it said, or said themselves, "If only I could have identified the fly they were taking . . ."

If this lament becomes a thing of the past, it may be due largely to John Goddard, whose indefatigable industry has given birth to *Trout Fly Recognition*. No longer need the fly of the moment remain shrouded in mystery, for every fly-fisherman possessed of this volume and a fly-net (though I personally use my cap) for catching insects should find no difficulty whatsoever in classifying and naming trout flies correctly, or at least as correctly as is necessary for fishing purposes.

It is a fact that almost every student of fly-fishing throughout history has eventually confessed to a belief in what I lightly term, from the name of the poet, Moses Browne, who wrote it, the Mosaic Code of dry fly fishing.

> When artful flies the angler would prepare,
> This task of all deserves his utmost care:
> Yet thus at large I venture to support,
> Nature best followed best secures the sport,
> Of flies the kinds, their seasons, and their breed,
> Their shapes, their hue, with nice observance heed.

This book will certainly help the fly-tyer, both the beginner and the old hand, to "heed with nice observance the shape and hue" of the flies he wishes to imitate; indeed, I doubt if there are any fly-fishers anywhere who would not acquire much knowledge and derive much benefit from it.

I recommend it also to those fly-fishermen whose experience leads them to reject the imitation theories of the majority, for the study, even to a limited degree, of aquatic insects is a fascinating one, and likely to enhance considerably the enjoyment of their visits to the waterside. I know of no better way to escape, however temporarily, from the pressures and strains of modern life than by delving a little into the wonderful life history of the beautiful insects we encounter at the rivers and lakes we fish.

The photographs which form the nuclei of this book were not at first intended for publication, and it was due only to the urgings of several of his colleagues to whom he showed them that the Author decided to expand his original intention, which was merely to provide photographs of insects in natural colour as guides for his own use when tying artificials during the close season, into its present form. This required a great expenditure of time, endurance, constant travel, and unremitting correspondence with experts in various fields of entomology.

Some interesting facts, not entirely connected with identification, came to light in the course of the intensive search for specimens, and, to the more advanced students of fishermen's insects, these facts are of no little importance. Thus, the manner in which the Sherry Spinner carries her egg ball can no longer be disputed; the belief that the Blue Winged Olive lives only in running water has been effectively shattered; and the unreliability of pedantic dogma relating to the colour, markings and size of specific species is clearly exposed.

In his chapter on Sedges, I believe John Goddard has made an attempt to clarify and identify species in a way and to a degree never before attempted. He is the only writer who has tackled the subject with the vigour it demands, and future generations of fly-fishers may well owe him a debt of immense gratitude.

John Veniard, who contributed the chapter on artificial fly dressing, needs little introduction to the world of fly-fishing. His connection with the firm that bears his name has linked him intimately with both the traditional and contemporary techniques of fly imitation and the usages of materials; his books on the subject have established him as an outstanding authority here and overseas. His contribution is a worthy suffix to a monumental manual.

London, S.W.1   DAVID JACQUES

AUTHOR'S PREFACE

This book has been produced in order to simplify the identification of the river-bred flies on which trout feed.

Most books on this subject which have appeared during the last century are either over-simplified or too complex, and in both cases this makes positive identification of any given fly very difficult.

It has, therefore, been decided to limit the flies described to the more common species. Flies that are seldom seen and those confined to a few local areas have been excluded. Wherever possible the flies are referred to by the popular English names that most fly-fishermen know and recognize.

The modern fly-fisher probably has a wider knowledge than his forebears due to the number of articles now published in the many angling journals; nevertheless it is surprising how many fly-fishers today lack even the most elementary knowledge of entomology. This is a pity as even a small knowledge of fly life can add much to the enjoyment one can find at the waterside while fishing.

The angler who is able to identify the fly that is hatching at any given time probably has the edge on the angler who lacks this ability, and this may make the difference between an empty creel and a satisfyingly full one.

Unfortunately, for the angler, trout rarely feed to any given pattern, and when several varieties of fly are drifting down over a feeding trout, the fish may well prefer one particular species and feed on this to the exclusion of all others. Furthermore, its choice may not be the fly which is coming down in the largest quantity. Obviously, the angler who can quickly and positively identify the particular fly on which the fish is feeding has a tremendous advantage over the angler whose only recourse is to go through every artificial in his box until by pure chance he finds an effective pattern.

A working knowledge of fly fishing entomology is much less difficult to acquire than might be imagined. It is unnecessary to make a detailed study of the subject in order to become reasonably proficient, and with this book as a guide it can be quite an absorbing pastime, adding interest to those dour days or periods when fish are inactive and not interested in the flies hatching out.

To start with, a small muslin dipping net about 4 inches across, which can be attached to the reverse end of your landing net handle, is invaluable as the flies can then be collected for closer examination as they hatch out. In the early days, a small pocket magnifying glass of about 8× is essential, as this will greatly help to make accurate identifications.

These items may be dispensed with as knowledge grows, for in many cases the location of hatches, time of day, and behaviour of various species of flies are helpful guides to identification. For a few species a magnifying glass is really essential at all times, and takes up so little space that it is well worth a place in one's kit.

It has been decided not to give specific details for the positive identification of all the various nymphs in this book, as accurate artificial representations are of far less importance to the fly fisher than the manner of presentation of a general pattern of artificial nymph.

Although this book has been produced to assist in identifying the flies on which fish feed, this knowledge is completely useless to the angler unless he has first learnt to use his tackle effectively and has perfected the right approach to his quarry. The angler who has learned to stalk and cast correctly and present his artificial fly in the correct manner to a feeding fish, even if it is the wrong pattern, will usually catch more fish than the angler who knows his entomology but handles his tackle badly. So the novice is strongly recommended to perfect his fishing technique before attempting to master fly identification.

For those fly fishermen who require more detailed knowledge than is available in this book, there are several excellent works. Four of these are particularly worthy of mention: *An Angler's Entomology*, by J. R. Harris; *Life in Lakes and Rivers*, by T. T. Macan and E. B. Worthington; *A Key to the Nymphs of the Ephemeroptera*, by T. T. Macan; and *A Revised Key to the Adults of Ephemeroptera*, by D. E. Kimmins. The latter two are obtainable from the Freshwater Biological Association, Ambleside.

ACKNOWLEDGEMENTS

I should like to take this opportunity of thanking all those who have assisted both directly and indirectly in so many ways, and without whose help this book could not have been completed. In particular I should like to mention the following:

Mr. David Jacques (Angling Editor for A. and C. Black Ltd.) for the immense amount of time and trouble he has taken in reading and where necessary correcting the final MS., and also for his invaluable advice and many helpful suggestions in the latter stages of its preparation.

Major Oliver Kite for encouragement, advice and assistance he unstintingly gave me in the early stages, and also for the very many hours he spent checking the original MS. for errors and inconsistencies before the final MS. was prepared.

Dr. Michael Wade of Risca for the many specimens and much valuable information he has supplied on flies in South Wales.

Dr. T. T. Macan, M.A., Ph.D., of the British Freshwater Biological Association.

Mr. Eric Horsfall Turner, Editor of the *Angler's Annual*, and Mr. Thomas Clegg, the well-known fly dresser, for their advice and for the many specimens they provided.

Mr. Ian Wood, Editor of *Trout and Salmon*, who was originally responsible for suggesting I undertake this book based on my colour photographs.

Apart from the above, I would also like to acknowledge the advice and help of the following: Mr. D. E. Kimmins and Mr. Peter Ward of the Dept. of Entomology of the British Museum, Mr. Alex Behrendt of Romsey, the late T. K. Wilson of Skipton, Mr. J. H. Evans of Sutton Coldfield, Brigadier E. N. Oldrey of Fighelden, Lieut.-Col. W. T. Sargeant, Mr. Len Cacutt, Capt. Terry Thomas, Mr. Barrie Welham, and my many friends of the Piscatorial Society.

In the preparation of this book many waters were visited in Scotland, Wales, the West Country, as well as in the South of England, and I should like to thank all those who kindly allowed me access to their waters to obtain specimens and study insect life, especially Mr. Alan Dalton of Over Wallop, John Presant of Reading, Dr. Jack Jones of Swansea, Lionel Sweet of Usk, and Mr. Kennedy Brown of the Bristol Waterworks Corporation.

Finally, I should like to pay a special tribute to my friends Cliff Henry and the late Frank Fortey who personally helped me in the arduous task of obtaining the many hundreds of specimens of flies I needed.

Fly with roof-shaped wings—Trichoptera

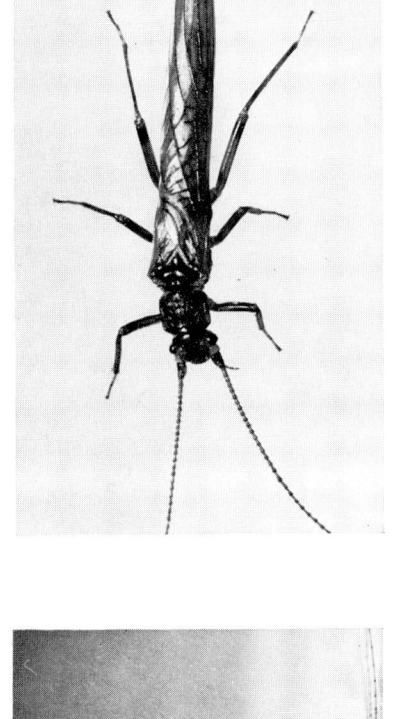

A hard-winged fly—Plecoptera

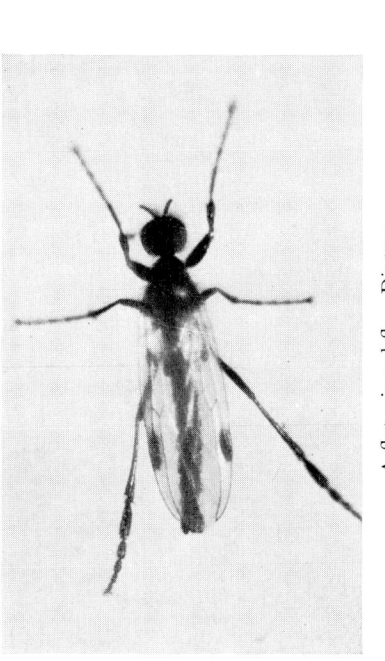

A flat-winged fly—Diptera

An upwinged fly—Ephemeroptera

FIG. 1. The Four Major Groups.

CHAPTER I

THE FOUR MAJOR GROUPS

CLASSIFICATION

The entomological Orders of Ephemeroptera, Diptera, Trichoptera and Plecoptera include most of the insects which are of importance to the fly-fisherman, and for simplicity these Orders can be described as Upwinged flies, Flat-winged flies, flies with Roof-shaped wings, and Hard-winged flies.

The first step in the identification of any given fly is to establish to which of the following groups the specimen belongs:

1. The Upwinged flies (Ephemeroptera)

All flies in this Order have a segmented body, two or three long tails and two large upright transparent or opaque wings. Almost all have two small hindwings.

2. The Flat-winged flies (Diptera)

The flies in this Order have two rather short transparent wings which lie flat along the top of the body (except *Tipulidae* spp., which hold wings at right angles to body). They have no tails and in general appearance, though not always in size, are somewhat similar in many cases to the common house-fly.

3. Flies with Roof-shaped wings (Trichoptera)

These flies have four wings, and when at rest the wings lie close along the body in an inverted V shape. The wings appear to be soft, as they are covered completely in very tiny hairs. They are without tails.

4. The Hard-winged flies (Plecoptera)

The wings, of which there are four, are long and rather narrow when the fly is at rest, and lie flat along and slightly over the body; they are hard and shiny or horny. Some of the larger species have tails. The wings of the male of many of these larger species are often very short and useless for flight.

A simplified explanation of the life cycle of flies in these Orders may be of interest, and as the first group is of more importance to the average fly-fisher than the other three, it is intended to give here a more detailed life history of the Upwinged flies than of those of other Orders.

THE UPWINGED FLIES (*EPHEMEROPTERA*)

Flies in this Order are known as the Upwinged flies. To quote a reference in many angling books, when a hatch of these occurs "they look like a fleet of miniature yachts sailing down the river".

The Egg

The egg stage is the first in the life cycle of flies of this Order. Eggs of most species are deposited by the adult female on the surface of the water, when they sink to the bottom and attach themselves naturally to weeds, stones, etc. Certain other species, such as some of the *Baëtis* genus, crawl down projecting weeds, stones, posts, etc., and deposit the eggs directly. The egg stage of development lasts anything from a matter of days to many months, according to species and time of year.

The Nymph

The nymphal stage is the second in the cycle, following the hatching out of the egg. This period may last from two to twelve months or even longer, according to species and the time of year the egg was laid. During this stage the nymph lives on or near the bottom. Some species hide in or on various weeds, others cling to stones and rocks, while certain other species actually burrow in the river bed.

It develops from a small nymph through progressively larger stages, each change accompanied by a moult. The growing nymph feeds mainly on decaying vegetable matter.

The Sub-imago or Dun

When the nymph is fully grown it is at last ready to change into the winged insect known as the sub-imago, or dun. This emergence takes place fairly rapidly. The mature nymph ascends to the surface, where the nymphal case splits (see Plate V. 70) and the fully winged fly emerges on to the surface film, resting on this film while its wings dry and it gathers strength to fly off.

On hot, dry days or in fast broken water this emergence is very rapid, as the wings dry very quickly and the insect flies off almost at once. The dun is slower to take wing on wet, damp, cold days. Certain species seem able to become airborne much faster than others. The dun stage usually lasts between twelve and thirty-six hours, much less in the case of flies belonging to the family Caënidae, where the transpositions from nymph to dun and dun to spinner may both take place in a matter of minutes.

The Imago or Spinner
The final stage in the life cycle of flies of this Order is after the change from the rather dull-looking dun to the beauty and perfection of the imago or spinner. It is during this final stage of its life, which may last only a matter of hours, that the fly engages in the procreative processes before it dies. The spinner can easily be distinguished from the dun as the latter is rather drab or dun-coloured, as its name implies. The wings are dull and opaque, and the trailing edges are usually lined with very fine hairs which can be clearly seen under a low-power magnifying glass. On the other hand, the spinner's body is very bright and shiny, with transparent shiny wings, now devoid of the fine hairs along the trailing edges; the tails and forelegs are also considerably longer.

The final stage of transposition from dun to spinner is accomplished fairly quickly in much the same way as the earlier stage from nymph to dun. This usually takes place in the bank herbage, anything from a few hours to a day or two after the appearance of the winged dun. The change from dun to spinner can actually be observed, with patience. If a dun can be successfully taken home alive, it should be deposited in a cool place, or better still kept overnight in an ice box or refrigerator, which usually results in slowing or halting the metamorphosis. On the following morning, place the fly in a warm spot in direct sunlight if possible, and within a fairly short time if you are fortunate the final transposition can be observed.

Duns at rest hold their wings upright and together, and usually the first indication of the forthcoming transposition is when the wings open out until they are spread wide apart. The dun sometimes spends several minutes in this position, quivering slightly from time to time, before the wings finally appear to fold slowly along the body. It is at this stage that the skin splits along the top of the thorax, and the fully adult fly or spinner emerges. (See Plate IV. 53 to 55 also Plate III. 36)

The cast-off skin retains the shape of the original insect, with the exception of the wings, which are too frail. (See Plate VIII. 109)

The fly now, as a spinner, is able to mate, and copulation usually takes place in flight. After this act, the spent male dies over land, but in some cases dies over and falls on to the water. The female, after extruding and depositing her eggs, usually dies in or on the water.

Finally, it should be explained that all flies in this Order have either two or three long tails, a segmented body, six legs and upright wings. Most of them have two large forewings and two smaller hindwings. In some cases the latter can be clearly seen, as in the Mayfly or Blue Winged Olive, but in other species, such as the Small Spurwing, they are so small that a magnifying glass is required to detect them. In a few species the hindwings are absent altogether.

THE FLAT-WINGED FLIES (*DIPTERA*)

This is an extremely large Order of insects, which includes all the true flies, such as house-flies, mosquitoes, dung-flies, crane-flies, etc., and is considerably larger than all the other three Orders put together. Despite this, relatively few of these flies are of significance to the river fly-fisherman.

Of the few flies in this Order that are of interest to anglers, the majority are aquatic. They can be divided into three main sections: Reed Smuts, Gnats and Midges. The latter, it should be mentioned, are very much more important to the lake than to the river angler. The vast majority of the two-winged flies in this Order are non-aquatic, and with a few notable exceptions (including some of the Gnats mentioned above) are of no importance to the angler.

Reed Smuts

The adult female insects in this section mainly lay their eggs in masses of jelly on protruding vegetation, stones, posts, etc., and may in some cases crawl down these projections and lay their eggs under water. The eggs usually hatch within a week or so, and small worm-like larvae emerge. These larvae moult at intervals, gradually increasing in size. Finally the larvae spin a cone-shaped case, usually attached to weeds, in which to pupate. When the adult fly is ready to emerge from its pupal case, it bursts out and ascends to the surface in a bubble of gas. (It is this feature combined with their small size which makes it so difficult to deceive a fish when it is feeding on these creatures.) As a result of this phenomenon the fly emerges at the surface quite dry, and usually immediately takes wing.

Gnats and Midges

Most of the flies in these two sections are larger than the very tiny

Reed Smuts of the previous one. The eggs of the Midges are usually laid by the adult females on the surface, in small or large clusters, forming rafts of eggs which attach themselves to weed or other projections.

The small larva which at length emerges swims to the river-bed, where it lives in the mud or silt. Eventually the larva transforms into a pupa which, unlike the previous species, is of the free-swimming variety, similar to the nymphs of the Olives. When the adult fly is ready to emerge the pupa ascends to the surface where it remains suspended for some little time in the surface film (depending on weather and water temperature). The winged insect then emerges and flies away.

All flies in this section are fully adult immediately after emerging from their pupal case, and pass only through the one-winged stage before they mate and die. It should be noted that some of the Gnats are terrestrial.

All flies in the Diptera Order are tailless, and have two short flat transparent wings similar to the common house-fly, which belongs to this same Order.

FLIES WITH ROOF-SHAPED WINGS (*TRICHOPTERA*)

The Caddis or Sedge-flies

This again is a fairly large Order of flies, about 190 different species having been recorded in these islands. The majority of these are of little interest to the angler; many are very small, while many others are very uncommon or locally distributed.

However, the few species of Sedge-flies that are common are very important indeed to anglers. They have roof-shaped wings (when at rest), four in number, which are covered with a layer of tiny fine hairs. These are sometimes difficult to see with the naked eye, but show up clearly with a magnifying glass. Superficially, they are similar to moths, but these latter have a covering of tiny scales on their wings instead of hairs. Also Sedges when at rest are considerably slimmer, particularly across the thorax.

Many Sedge-flies have particularly long antennae, in some cases over twice the length of their bodies. The mouthparts include two pairs of jointed protuberances called palps, which are set one behind the other. The front ones, the "maxillary palps", sometimes vary between male and female, and in a few families the male palps consist of three to four segments and the female's of five segments. In most families, however, both male and female palps consist of five segments.

The female Sedge-fly lays her eggs in one of three ways according to the species:

(a) On the water surface.
(b) By crawling under water and depositing the eggs.
(c) On herbage or vegetation overhanging the water.

The several hundred eggs are often laid in one gelatinous mass which adheres to any vegetation with which it comes in contact.

There are four well-marked stages in the life of the Sedge-fly: egg, larva, pupa, and the adult winged fly.

The Eggs

These begin to hatch into larvae after ten to twelve days and most species construct cases to protect their relatively soft bodies.

The Caddis Larvae

These are a familiar sight to most anglers, as practically every clump of weed pulled from the river bed will have its attendant host of Caddis cases in various shapes and sizes. In days gone by these used to be a popular bait with bottom fishermen, and even today are still much used in Ireland. The larvae form their tubular cases (see Plate V. 67) from all manner of debris from the bed of the river or lake, including particles of gravel, shells, bits of leaf, decaying vegetation and even small sticks. In fast flowing water the larvae usually choose heavy materials such as small stones, gravel, etc., to help prevent them being swept away by the current. On the other hand, the species found in still water tend to use semi-buoyant materials which facilitate movement.

The majority of flies in this Order form cases as described, but a few free-swimming species (see Plate V. 66) form shelters on the undersides of stones or plants. After the larva is fully developed, it pupates within its case or shelter, and after a period varying from days to weeks, according to species and the time of year, during which it does not eat, the pupa emerges. At this stage it is still enclosed in its pupal envelope and is equipped with a pair of powerful paddle-like legs which enable it to swim to the surface, or to climb up plant or weed stems above water where it hatches out into the fully adult fly. In some species the pupa swims directly to the surface, where it immediately hatches out. Large Cinnamon Sedges are typical of this latter group, and many anglers will be familiar with the disturbance they cause as they hatch on the surface and skitter about in a frantic effort to become airborne.

Sedge-flies in the winged state, perhaps because they are able to absorb liquid, live for a much longer period than flies of the Ephemeroptera group.

Finally, it should be noted that most of these flies mate while at rest, and therefore it is usually the female of the species that is sometimes taken by the trout when she returns to lay her eggs.

THE HARD-WINGED FLIES (*PLECOPTERA*)

Commonly called the Stoneflies, this is a small Order of flies consisting of a little over thirty different species. Nevertheless, where they occur in abundance, they are important to anglers. This is mainly in the North Country rivers that have gravelly or stony beds. They are of course found in some of our southern rivers, including many of the chalk streams, but seldom in sufficient quantities to be of interest to anglers.

The flies in this group vary considerably in size, from the large Stonefly, which is well over an inch long, to the small Willow-fly, and even smaller Needle-fly, the smallest fly in the Order. They all have four wings and when at rest the wings are long and narrow and lie close to the body. These wings are hard and shiny, prominently veined and considerably longer than the body of the fly, except in some males as mentioned earlier. They are poor fliers and in flight appear considerably larger than they really are. The adult fly is, like its nymph counterpart, active on its feet.

The life span of the flies in this Order varies considerably, the average being about a year, but in some cases lasting as long as two to three years.

The Egg

The time taken for the eggs to hatch out, again, varies from a few days to many weeks. The nymphs of Stoneflies are very robust and active creatures. Many anglers will be familiar with some of the larger nymphs in this group which are popularly known as creepers and in some circles are much in demand as bait for bottom fishing.

These nymphs can easily be distinguished from the nymphs of the Ephemeroptera as they have only two tails, and the two antennae on the head are quite long. In the latter stage of the life of these nymphs the wing cases are quite prominent. When the adult winged fly is ready to emerge from its nymphal case, the mature nymph crawls on to dry land where the transition takes place. The life of the Stoneflies in the winged state can be anything from a few days to a few weeks according to species, and they never wander very far from water, and mate at rest.

In many of the larger flies in this Order, the wings of the male are little better than stumps, and these insects are quite incapable of flight.

The females return to the water to lay their eggs, and the method of doing so varies a little according to the species. So far as can be ascertained, however, they all lay their eggs actually on the surface. With some of the larger species, they will alight on the water and release the eggs while swimming or fluttering along. In doing so they create quite a disturbance, and make an attractive target for any hungry trout in the vicinity.

Other species fly out over the water and literally fall down on the surface, releasing all their eggs in one mass, while some dip up and down on the water in a similar manner to some of the Upwing spinners, depositing a few eggs at a time. The eggs when released by the female on the surface sink to the bottom where they attach themselves to stones, rocks, etc.

N.B. In the following chapters it will be noted that the emphasis on identification heavily favours the Ephemeroptera. The reason is that most of the flies in this Order hatch in very large regular numbers, floating on the surface as duns and returning to the water as spent spinners. Therefore correct identification is more important in this than in other Orders.

However, it is also appreciated that in certain areas this Order of flies is of less importance and I ask for the indulgence of anglers in those areas for the sake of the majority.

The Orders or Classes of interest to fly-fishermen are as follows. They are listed according to importance.

Ephemeroptera	Upwinged flies.
Trichoptera	Sedge-flies.
Plecoptera	Stoneflies.
Diptera	Midges, Gnats, etc.
Crustacea	Shrimps, Water-louse, etc.
Megaloptera	Alder-flies.
Hemiptera	Water-bugs, Corixids, etc.
Coleoptera	Beetles.
Arachnidae	Spiders.
Lepidoptera	Moths, etc.

Neuroptera	..	..	Lacewings, etc.
Orthoptera	..	..	Grasshoppers.
Hymenoptera	..	..	Wasps, Ants, etc.
Odonata	..	..	Dragonflies and Damselflies.

All the above with the exception of Crustacea and Arachnidae belong to the Class Insecta.

For the fly-fisher intending to make a superficial study of entomology, it is important to note that all insects are divided into Orders, which in turn are divided into Families, and then into Genera, and finally into the Species. The following table shows how two different Orders of insects are classified.

	MAYFLY	BLACK GNAT
CLASS	*INSECTA*	*INSECTA*
ORDER	*EPHEMEROPTERA*	*DIPTERA*
FAMILY	*EPHEMERIDAE*	*BIBIONIDAE*
GENUS	*EPHEMERA*	*BIBIO*
SPECIES	*DANICA*	*JOHANNIS*

CHAPTER II

THE UPWINGED FLIES (EPHEMEROPTERA)

WITH DETAILED NAME KEY

In the opening stages of Chapter I, the life cycle of this Order has been dealt with fairly concisely, and it is not felt necessary to elaborate further for fear of confusing the reader. However, there are many other features and facts about this interesting and fascinating group of flies which, if explained in a straightforward and simple manner, can be of considerable interest to the angler. It is therefore proposed to deal with them step by step as we go along.

It should first be explained that although the winged flies in this group are important from the angler's or fly-fisherman's point of view, they are not necessarily of so much importance from the trout's point of view, as they may not form a substantial part of its diet. Trout feed on a great variety of food; snails, shrimps, crayfish, caddis and the larvae of other aquatic insects such as Smuts and Midges. In addition, trout are not averse to feeding on the smaller specimens of their own kind. Minnows, Sticklebacks and the fry of other fish also form an important part of their diet in many waters, as do the nymphs of the two Orders of flies. On occasions, also, both trout and grayling are fortunate in coming across such tit-bits as tadpoles, caterpillars, beetles, wasps, and various other land-bred insects which find their way on to or into the water. Therefore with this large variety of food to choose from, it is sometimes surprising that trout ever bother to rise at all to small Ephemeropteran duns or spinners. It is very fortunate for the fly-fisher that trout and grayling feed regularly on surface flies; they do so probably because these flies are conveyed by the current to the positions where the fish lie, thus enabling the latter to feed heartily with the expenditure of minimum effort. Another possibility, of course, is that adult flies supply a deficiency in the trout's diet and are necessary to its well-being, or perhaps it looks upon them as either an *hors-d'oeuvre* or a *dessert*.

Let us now consider the physical structure or characteristics of flies in this Order:

The Head
This is often wider than its length, the major portion consisting of the eye structure; there is no mouth as such, the duns being incapable of eating or drinking. This may account for their very short life span in the final winged state. The head also has a small pair of antennae like miniature horns, which are usually plainly visible.

The Eyes
The head carries a pair of large compound eyes or oculi, one on each side, and in most cases the eyes of the male are considerably enlarged and extend over the top of the head. These are probably provided by nature to enable it to locate the female of the species, which usually approach the male swarms from a higher level.

The Thorax
The head is connected by a short neck to the thorax of the fly, which is made up of three segments. The middle segment is by far the largest, and carries the forewings and the middle pair of legs. The front segment carries the forelegs and the rear segment the hindwings (if any) and rear pair of legs.

The Abdomen
This, the main body of the fly, is divided into ten distinct segments. The last segment carries the tails and the ninth segment carries the claspers of the male fly. The top of the abdomen will be referred to as the dorsum, and the under surface as the venter.

The Tails
They are usually long and slender. In some species there are two and in others three. In many cases these tails are very short when the dun is freshly hatched, but they rapidly lengthen with age. In the spinner stage the tails are usually considerably longer.

The Wings
As previously explained, these are fairly large with extensive veining and are always carried upright. Most of the flies in this group have hindwings of various sizes and shapes, but four species have no hindwings at all.

The Legs
These are six in number, the front or anterior legs being normally longer than the remaining four, which are referred to as the median

and posterior legs respectively. The anterior legs of the males are usually a little longer than those of their female opposites; in the spinner stage they are often longer still. Each leg consists of five main joints; the most important of these are the femur (thigh joint), tibia (shin joint) and tarsus (foot).

So much for the physical structure of these flies, but before we proceed further let us consider from the angler's point of view the simplest way of determining the sexes. This can be achieved by two methods, the first of which concerns the eyes. In the male these are usually very prominent, and to clarify this the following sketches

Fig. 2. Eyes of a typical female.

Fig. 3. Eyes of a typical male.

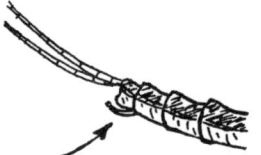

Fig. 4. Showing claspers of male fly viewed from the side.

Fig. 5. Showing close up of male claspers viewed from under.

should be studied. The first sketch shows the eyes of a typical female, and the second those of the male. (See Figs. 2 and 3.)

It must be pointed out, however, that to identify the males by the eye characteristics alone is not conclusive. In certain of the larger families of flies in this Order, this difference in structure is not so apparent. Therefore for conclusive evidence the second method is infallible. All male flies in this Order have a pair of forceps or claspers to enable the male fly to attach itself to its chosen mate during copulation. These are situated underneath the fly just behind the tail on the ninth segment (see Figs. 4 and 5). They can usually be seen with the naked eye, but a small magnifying glass will show them quite clearly. In conclusion, it should be noted that in nearly all cases the male fly is smaller than the female of the same species.

DIFFERENCE IN SIZE AND COLOUR

It is probably not generally realized how much variation it is possible to find in the colour and size of Ephemeropteran flies of the same species. This, of course, can greatly increase the difficulties of positive identification, as we automatically tend to recognize certain flies by their general size and colour. Therefore these variations must be kept very much in mind. But fortunately these variations seem to occur only occasionally and it is possible to work to average sizes and colours in most cases with reasonably accurate results. Let us first discuss the question of colour.

Colour

As a general rule the temperature of the air is thought to affect the coloration; the colder the day, the darker the colour of the fly. To follow this temperature effect further, flies that hatch in the spring or early in the year are darker than flies of the same species that hatch during the summer. Then, when the weather begins to get colder in the autumn, the colour is again darker. Another point also to be borne in mind is that the male is usually darker than the female. Apart from the above, the colours of the same species of fly can vary in different localities. However, where a colour variation does occur, it is more a lightening or darkening of the basic colour. One notable exception is the Blue Winged Olive. When this fly first appears in early June the body colour of the female is sometimes quite a bright olive green, but in October the colour is often quite different, being a distinct shade of rusty brown. Little has been written on the subject of size and coloration in the past, and during the preparation of this book hundreds of colour photographs have been taken and many thousands of specimens have been examined, and it has been during this time that these variations have been noticed.

Having dealt briefly with colour as applicable to the duns, let us now consider colour in respect of spinners. With these the variation in colour seems to be very much greater, and is much more noticeable in certain species than in others. Whether or not air temperature plays much part in coloration in the spinners is not definitely known, but it is generally accepted that the older the spinner the darker the colour. This is particularly noticeable in the females, and they seem to darken very rapidly after copulation. Although this progressive darkening with age seems fairly constant, it does not bear any relation to the final colour of any particular spinner, as this seems to depend more on the original colour of the spinner when it changes from the dun stage. The following two examples may help to illustrate this point:

(*a*) The female spinner of the Medium Olive (*Baëtis vernus*)—this often first appears with a greenish-brown body, and, when it is spent, with a dark brownish body, tinged red. The more common coloration, however, is a brownish body to start with, finishing up in the spent stage with a distinctly reddish body. It will be appreciated that the variation here between these two varieties is not so great.

(*b*) The female (Sherry) spinner of the Blue Winged Olive seems to be subject to a very large variation, and on more than one occasion I have seem swarms of these spinners ranging in colour from a pale olive-brown to a deep lobster-red. So far as I have been able to ascertain the exact cause of this variation in colour is unknown. Dr. Michael Wade suggests that mutant factors might ensure a mixture of colours in each generation so that against different coloured backgrounds some will be less conspicuous to predators and thus more likely to survive.

Difference in Size

Fortunately from the point of view of identification this variation in size seems to be the exception to the rule, and it is only the odd fly here and there that is either very considerably smaller or larger than the norm. If there is a variation it is more likely that the specimens will be smaller than average rather than larger. It is quite common to encounter complete hatches of the various species in this Order that vary in size slightly. Once again one can only theorize on the reason for this, and in all probability it is simply a lack or an abundance of food during their early stages of life under water.

It must also be borne in mind that a small variation in size can be accounted for by a natural tendency of all living creatures to differ in this respect. To give a comparison of this as against human beings, half a millimetre difference in the length of a B.W.O. (which is quite common) would be roughly equivalent to a difference in a man's height of about eight inches. Therefore a variation of up to one millimetre or more would not be abnormal in a fly.

MATING AND FLIGHT

Mating

Although this was dealt with superficially in Chapter I it is felt that a little more detail would be of general interest. In most cases mating takes place in the air. Once a male has located a female of the species, he approaches her from underneath and with the aid of his forceps (claspers) he attaches himself to her. Copulation then immediately takes place, during which the female endeavours to keep both herself and her mate airborne. The pair may slowly sink towards the ground, but in most cases copulation has been effected before they actually reach it. After mating has taken place the female will usually fly out over the water in preparation for laying her eggs and the male will return to its swarm, possibly preparatory to mating with other females. Eventually the males when fully spent will return inland where they die. Few males actually fall spent on the water in any quantity. However, there are exceptions to this, and three that immediately come to mind are the Small Spurwing, the Iron Blue and the Blue Winged Olive. The males in these cases sometimes fall spent on the water in considerable numbers. The actual reason for this seems difficult to pinpoint, but from observations I have made it would appear that these particular species often mate over the water, whereas most of the other species in this Order usually copulate over land.

Flight

The male spinners of the Ephemeroptera usually form themselves into swarms of varying size. Each swarm may be composed of several hundred or even several thousand individuals. Each different species seems to have its own favourite area or locality for swarming. For instance, the Pale Evening spinner is usually found swarming along the edges of rivers and seldom inland. On the other hand, the Small Red spinner is often to be found inland, sometimes as much as sixty or seventy yards from the water. Also, apart from the location of the individual swarms, the actual flight pattern or behaviour of flies of the different species varies. From the above it will be apparent that with experience the keen observer may eventually be able to recognize and identify some of the different species of spinners while in flight or by their location. For the angler, a little knowledge of the behaviour of these spinner swarms can be important. Certain species fall spent on the water earlier than others, and as most swarms form late in the afternoon and seldom fall spent till early or late evening, the correct identification of spinners in a swarm can assist the angler to be in the right place at the right time with the right artificial. Spinners, however, are very susceptible to a change in weather conditions, especially to a drop in temperature, and indeed on cold days they are often absent. Also if a cold wind blows up in the evening, bringing about a temperature drop, the swarms which might have formed throughout the afternoon will suddenly disappear to await a more favourable opportunity. This may not arise until the next day.

On some rivers the banks vary between open meadows and sheltered woodland. In inclement weather, the exposed open meadows will be deserted, and the spinner swarms may then often be located in the sheltered lee of the trees and shrubs.

DETAILED NAME KEY FOR THE EPHEMEROPTERA GROUP

CURRENT ANGLING NAME	OLD ANGLING NAME	ENTOMOLOGISTS' NAME	POPULAR NAME FOR FEMALE SPINNER
MAYFLY	Green Drake	*Ephemera danica* *Ephemera vulgata*	Spent Gnat.
LARGE DARK OLIVE	Large Spring Olive or Blue dun	*Baëtis rhodani*	Large Dark Olive spinner or Large Red spinner.
IRON BLUE	Iron Blue	*Baëtis pumilus* or *Baëtis niger*	Iron Blue spinner or Little Claret spinner.
MEDIUM OLIVE	Medium Olive	*Baëtis vernus* or *Baëtis tenax* or *Baëtis buceratus*	Medium Olive spinner or Red Spinner.
SMALL DARK OLIVE	Summer Olive, July dun or Pale Watery	*Baëtis scambus*	Small Red spinner or Small Dark Olive spinner.
PALE WATERY	Pale Watery	*Baëtis bioculatus*	Golden spinner or Pale Watery spinner.
DARK OLIVE	—	*Baëtis atrebatinus*	Dark Olive spinner.
YELLOW EVENING DUN	Yellow Evening dun	*Ephemerella notata*	Yellow Evening spinner.
BLUE WINGED OLIVE	Blue-winged Olive	*Ephemerella ignita*	Sherry spinner.
SMALL SPURWING	Little Sky-blue or Pale Watery	*Centroptilum luteolum*	Little Amber spinner.
LARGE SPURWING	Blue-winged Pale Watery or Pale Watery	*Centroptilum pennulatum*	Large Amber spinner.
PALE EVENING DUN	—	*Procloëon pseudorufulum*	Pale Evening spinner.
POND OLIVE	Pond Olive	*Cloëon dipterum*	Pond Olive spinner or Apricot spinner.
MARCH BROWN	March Brown	*Rhithrogena haarupi*	March Brown spinner.
OLIVE UPRIGHT	Olive Upright	*Rhithrogena semicolorata*	Yellow Upright.
YELLOW MAY DUN	Little Yellow May dun or Yellow Hawk	*Heptagenia sulphurea*	Yellow May spinner.
DUSKY YELLOWSTREAK	Dark dun	*Heptagenia lateralis*	Dusky Yellowstreak spinner.
TURKEY BROWN	Turkey Brown	*Paraleptophlebia submarginata*	Turkey Brown spinner.

CURRENT ANGLING NAME	OLD ANGLING NAME	ENTOMOLOGISTS' NAME	POPULAR NAME FOR FEMALE SPINNER
PURPLE DUN	—	*Paraleptophlebia cincta*	Purple spinner.
DITCH DUN	—	*Habrophlebia fusca*	Ditch spinner.
CLARET DUN	Claret dun	*Leptophlebia vespertina*	Claret spinner.
SEPIA DUN	—	*Leptophlebia marginata*	Sepia spinner.
AUTUMN DUN	August dun	*Ecdyonurus dispar*	Autumn spinner.
LARGE BROOK DUN	—	*Ecdyonurus torrentis*	Large Brook spinner.
LARGE GREEN DUN	Large Green dun	*Ecdyonurus insignis*	Large Green spinner.
LATE MARCH BROWN	Late March Brown or False March Brown	*Ecdyonurus venosus*	Great Red spinner.
CAËNIS OR BROADWING	Angler's Curse or White Midge	*Caënis* and *Brachycercus* spp.	*Caënis* spinner or Broadwing spinner.

With regard to the above list, it should be explained that in the early days of fly-fishing most of the species of the *Baëtis* genus, apart from the Iron Blue (*Baëtis pumilus* or *niger*), were referred to generally as Olive duns. In addition the two flies from the *Centroptilum* genus and one of the flies from the *Baëtis* genus, the Pale Watery (*Baëtis bioculatus*), were all referred to as Pale Watery. This has led to a certain amount of confusion in the past, and therefore in this name key and the identification keys I have designated them in accordance with contemporary thought. It must be pointed out, however, that the above insects are very similar in general appearance and are far from easy to identify positively. In fact in one or two cases the difference is so slight that it is very doubtful indeed whether correct identification is of any help at all to the angler.

Finally it must be emphasized, as mentioned in the introduction, that the above list does not include all the insects in the Ephemeroptera Order. Those flies that are seldom seen or confined only to local areas have been omitted for fear that their inclusion might lead to some confusion.

APPENDIX TO CHAPTER II—*THE BAËTIDAE*

Before proceeding to the next chapter, it has been thought advisable to give some brief details and information on the Baëtidae family of Olives, as of all the Upwinged flies these are undoubtedly the most difficult to identify precisely. They are also dealt with in the following chapters, but as a rough guide the following points should be noted.

The *Baëtis* genus are a member of the Family Baëtidae, as are also the genera *Centroptilum*, *Cloëon* and *Procloëon*. As some of the generic species within this family differ only little from each other in size and colour, it is almost impossible to identify them positively by these characteristics alone. Therefore it is necessary to indicate certain physical differences in order to enable one to subdivide the Baëtidae family into their correct genera and species.

From the fishing viewpoint the importance of correct identification within these groups is in any case of dubious value. A range of patterns from size 000 up to 1 in a pale dressing to represent the Pale Watery, Pale Evening dun and Spurwings, and similar sizes tied with dark olive and medium olive dressings to represent the Olives and Iron Blue, will generally suffice.

However, in certain cases it can be helpful to identify the naturals, and as this is often a lot simpler than would at first appear, it is well worth a little time and trouble to master the processes of identification.

First of all let us look at the complete list of Upwinged duns that come in this group. It will be noted that some species in this group have been bracketed together, the difference between them being very slight so far as the angler is concerned.

Baëtis rhodani	The Large Dark Olive
Baëtis atrebatinus	The Dark Olive
Baëtis vernus	⎫
Baëtis tenax	⎬ The Medium Olive
Baëtis buceratus	⎭
Baëtis pumilus	⎫ The Iron Blue
Baëtis niger	⎭
Baëtis scambus	The Small Dark Olive
Baëtis bioculatus	The Pale Watery
Centroptilum luteolum	The Small Spurwing
Centroptilum pennulatum	The Large Spurwing
Procloëon pseudorufulum	The Pale Evening dun
Cloëon dipterum	The Pond Olive

Baëtis tenax has been listed above although there is doubt as to whether this actually exists as a separate species, but as it has been grouped with the other two Medium Olives, this is of little consequence. *Cloëon simile* has not been included above as it is a species found mainly in the deeper ponds and lakes, and therefore does not come within the scope of this **book**.

THE UPWINGED FLIES (EPHEMEROPTERA) 33

The following table has been prepared giving the various physical characteristics of the flies in the above list. This should enable the reader to identify most of them correctly.

All the flies in this group have *two tails*. Some of them have

1. Single or double marginal intercalary veins along the trailing edge of the forewings. See Figs. 13 and 14 (p. 40)
2. Narrow or oval hindwings. See Figs. 13 and 14 (p. 40)
3. No hindwings at all.

These form the basis of the following table.

IDENTIFICATION CHART FOR THE *BAËTIDAE* FAMILY OF DUNS

SPECIES	SIZE	COLOUR OF EYES OF MALE	SHAPE OF SMALL HINDWING	INTER-CALARY VEINS	REMARKS
LARGE DARK OLIVE (*B. rhodani*)	MEDIUM-LARGE	DULL BRICK-RED	SMALL OVAL WITH SPUR (Fig. 18)	DOUBLE	These two are difficult to tell apart except for difference of hindwing. Both have dark olive-brown bodies.
DARK OLIVE (*B. atrebatinus*)	MEDIUM	PALE RED-BROWN	SMALL OVAL BUT NO SPUR (Fig. 19)	DOUBLE	
MEDIUM OLIVE (*B. vernus, tenax* or *buceratus*)	MEDIUM	DULL RED-BROWN	SMALL OVAL WITH SPUR (Figs. 20 & 23)	DOUBLE	Body colour medium olive.
IRON BLUE (*B. pumilus* or *niger*)	SMALL	DULL RED-BROWN	SMALL OVAL WITH SPUR (Figs. 16 & 17)	DOUBLE	Body very dark brown-olive with dull grey-blue wings.
SMALL DARK OLIVE (*B. scambus*)	SMALL TO VERY SMALL	DULL ORANGE-RED	SMALL OVAL WITH SPUR (Fig. 21)	DOUBLE	The females of these two species are anatomically almost identical. Body colour is grey-olive and pale grey-olive respectively.
PALE WATERY (*B. bioculatus*)	SMALL	YELLOW	SMALL OVAL WITH SPUR (Fig. 22)	DOUBLE	
SMALL SPURWING (*C. luteolum*)	MEDIUM TO SMALL	ORANGE-RED	VERY SMALL NARROW WITH APEX POINTED AND PROMINENT SPUR (Fig. 25)	SINGLE	Pale olive body.

SPECIES	SIZE	COLOUR OF EYES OF MALE	SHAPE OF SMALL HINDWING	INTER-CALARY VEINS	REMARKS
LARGE SPURWING (*C. pennulatum*)	MEDIUM LARGE	DULL ORANGE	VERY SMALL NARROW WITH APEX ROUNDED AND PROMINENT SPUR (Fig. 26)	SINGLE	Very pale olive-grey body with dark blue-grey wings.
PALE EVENING DUN (*P. pseudorufulum*)	MEDIUM	DULL YELLOW	NO HIND-WINGS	SINGLE	Very pale straw body.
POND OLIVE (*C. dipterum*)	MEDIUM TO MEDIUM-LARGE	DULL ORANGE-RED	NO HIND-WINGS	SINGLE	Dark brown-olive body.

Finally it should be pointed out that in the male flies of all the above species, and indeed in all the male flies of the Ephemeroptera Order, identification is assisted by the shape and characteristics of the male genitalia. However, as these can only be studied with the aid of a microscope, they are not recommended as a means of identification in this book.

N.B. It should be noted that the true colour of the eyes of the male duns may not be apparent when freshly hatched. It is therefore advisable to let a short period elapse before examining freshly hatched males.

CHAPTER III

TWO SIMPLE IDENTIFICATION KEYS FOR THE DUNS AND SPINNERS OF THE EPHEMEROPTERA ORDER (THE UPWINGED FLIES)

KEY NO. I (FOR THE ANGLER/ENTOMOLOGIST)

KEYS TO THE EPHEMEROPTERA   THE UPWINGED FLIES

These simple keys have been evolved to enable the angler interested in entomology to identify fairly quickly and accurately most of the duns or spinners. The keys are split into size and other groups. It is only necessary to compare the specimen to be identified in the first instance with the main group, A, B, C, etc. As soon as the specimen has been identified with a main group, and not before, it can be followed down through the size-groups, etc., until the field is narrowed down to a particular fly.

In some cases, however, it may be possible to narrow it down only to two or three possible flies, and reference will then have to be made to the next chapter for positive identification.

As far as possible, distinct physical characteristics are used as the basis of these keys. In the key to the duns, the wings are often of help in identification, and reference is made to these where applicable.

In the key to the spinners, the wings are of little help in some cases and other features are therefore included.

It is dangerous to rely too much on size or coloration, but average sizes can be of some assistance in identification. For these keys, and also throughout the rest of the book, the following size references will be used. (See Page 36.)

Before you refer your specimen to the key, it is of course first necessary to decide whether it is a dun or a spinner.

The Dun
The colours are usually rather drab and the wings are often opaque

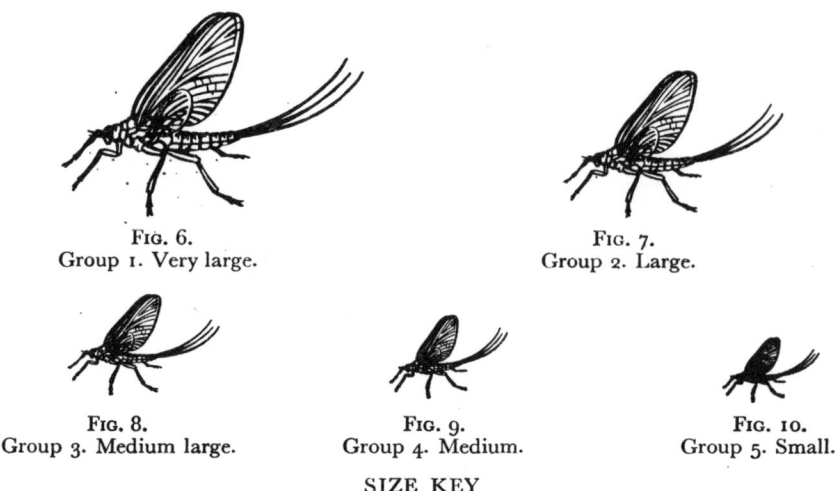

Fig. 6.
Group 1. Very large.

Fig. 7.
Group 2. Large.

Fig. 8.
Group 3. Medium large.

Fig. 9.
Group 4. Medium.

Fig. 10.
Group 5. Small.

SIZE KEY

or coloured. The trailing edge of each forewing is fringed with very fine hairs.

The spinner

The colours are usually brighter or shiny. The tails are often considerably longer than in the dun, especially in the case of males. The wings are nearly always transparent and shiny, and no longer have the fine hairs along the trailing edge (except in the case of *Caënis* spinners).

It may be necessary in some cases to use a pocket magnifier.

IMPORTANT. REFER TO MAJOR LETTER GROUP HEADINGS *ONLY* TO START WITH UNTIL SPECIMEN IS IDENTIFIED TO CORRECT GROUP. IT CAN THEN BE NARROWED DOWN THROUGH SIZE-GROUPS, ETC.

Remember that the Keys are based on certain features. They are:

(1) Number of tails (two or three).
(2) Number of wings (two or four).
(3) Shape of hindwings, if any.
(4) Venation of wings.

These are the main points to look for when a quick identification is required. Occasionally, the genitalia will need microscopic examination for positive identity, but this can be left until the angler is more proficient; it is not, however, dealt with in this book.

The reader must be on his guard against two things. First, occasionally an insect loses a tail, and consequently a three-tailed fly can look as if it possesses only two. Second, in one or two species the hindwings are very small and can easily be overlooked.

The following two examples are quoted:
Imagine you are at the waterside and have just caught a fly.

Example No. 1
1. Is it a spinner or a dun? Answer—a dun.
2. Now look at group headings ONLY. Starting with Group A, "Duns with upright hindwings and three tails". It has upright hindwings and three tails, so it comes in this group.
    Having now established which group it is in, ask:
3. What size? (refer to separate size key) Answer—medium-large, so it comes in size-group 3. As this group contains several species, we look for further information in an effort to narrow the field. Under "Identifying Features" we find the colour of the wings on some species is mentioned. Our specimen has dark blue-grey wings with an olive-green body. So it must be a BLUE WINGED OLIVE.

Example No. 2
1. Is it a spinner or a dun? Answer—a dun.
2. How many tails? Answer—two. This takes us to groups C, D, E, F, and G.
3. How many wings? Answer—four. This narrows the search to groups C, D, E and F.
4. What is the shape of the hindwings? Answer "oval". This leaves us definitely at group E.
5. What size? (refer to separate size key) Answer—small.

In size-group 5, "Small", we have three flies. In an effort to narrow it down still further, we see under "Identifying Features" one of these three flies is mentioned, i.e. "The Iron Blue". Your specimen does not have grey-blue wings, so it is not this fly. You are now left with two flies. The Pale Watery and the Small Dark Olive. In this case, to find out which of these two your specimen is, it will be necessary to refer to the next chapter, where specific details of all flies in the Ephemeroptera Order are listed.

It is recommended that even where these keys bring you down to a specific fly, it is as well to check it against the full description in the next chapter. When checking the specimen you are trying to identify against the size key, it should be remembered that the male fly is a little smaller than the female in most species of Ephemeroptera.

THE UPWINGED FLIES
KEY 1. THE DUNS (MALE OR FEMALE)
DUNS WITH THREE TAILS

GROUP A. Duns with upright hindwings (see Fig. 11) and three tails.

FIG. 11. Showing forewing and upright hindwing with slight costal projection.

FIG. 12. Showing upright hindwing without costal projection.

Size-group 1. Very large.   Mayfly.
           3. Medium-large. Blue Winged Olive, Yellow Evening dun, Sepia dun, Turkey Brown and Claret dun.
           4. Medium.   Purple dun, Ditch dun.

Identification features

Blue Winged Olive	wings dark blue-grey, body olive-green to rusty brown-olive.
Yellow Evening dun	wings pale yellow, body pale yellow.
Ditch dun	upright hindwings *with prominent costal projection.* (See Fig. 24)
Sepia dun	upright hindwings *without costal projection.* (See Fig. 12) Wings pale fawn, heavily veined.
Turkey Brown	wings mottled, fawn with blackish veining, pale area in centre, body dark blackish brown. Upright hindwings *without costal projection.* (See Fig. 12)
Claret dun	fore wings very dark grey. Pale buff upright hindwings *without costal projection.* (See Fig. 12) Body dark brown with claret tinge.
Purple dun	blackish grey wings, body dark brown-purple. Upright hindwings *without costal projection.* (See Fig. 12)

GROUP B. Duns with no hindwings and three tails.

Size-group 5. Small or very small. *Caënis*.

Identification features
Wings short and broad, body dull cream.

DUNS WITH TWO TAILS—ALL REMAINING GROUPS

GROUP C. Duns with fawn mottled wings, upright hindwings (see Fig. 11) and two tails.

Size-group 2. Large. Autumn dun, Large Green dun, March Brown, Late March Brown, Large Brook dun.

Identification features

Large Brook dun	pale fawn mottled wings, banded with dark markings and clear area in centre, leading edge yellowish. Drab olive-brown body, with red-brown diagonal bands along sides.
Large Green dun	pale fawn mottled wings with a distinct pale area in centre of forewings. Body dark olive-green with diagonal brown bands along sides.
March Brown	pale fawn mottled wings with distinct pale area in centre of forewings. Body dull brown. Brown streak present in middle of each femur (top leg section).
Late March Brown	fawn mottled wings. Dull brown body. Mid-season species.
Autumn dun	pale fawn mottled wings. Pale olive-brown body. Late season species.

GROUP D. Duns with yellow or grey wings, upright hindwings (see Fig. 11) and two tails.

Size-group 3. Medium-large. Yellow May dun, Olive Upright, Dusky Yellowstreak.

Identification features

Yellow May dun	wings pale yellow, body pale yellow.
Olive Upright	wings dark blue-grey, body grey olive-brown. Also has a distinct brown streak in middle of each (top leg) femur section.
Dusky Yellowstreak	wings very dark grey, body dark greyish brown with a yellow streak in front of forewing roots each side of thorax.

GROUP E. Duns with double marginal (paired) intercalary veins, small oval-shaped hindwings and two tails. (See Fig. 13)

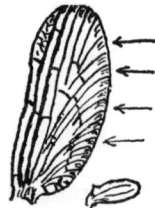

FIG. 13. Showing paired intercalary veins and oval hindwing—with spur shaped costal projection.

Size-group 3. Medium-large. Large Dark Olive.
            4. Medium. Medium Olive, Dark Olive.
            5. Small. Pale Watery, Small Dark Olive, Iron Blue.

Identification features
   Iron Blue   wings dull grey-blue, body very dark brown-olive.
   Large Dark Olive wings pale grey, body dark olive-brown.
   Dark Olive   no spur on hindwing. (See Fig. 19).
   The male duns of all above have red-brown eyes, except the Pale Watery which has yellow eyes.

GROUP F. Duns with single intercalary veins, very small narrow hindwings and two tails. (See Fig. 14)

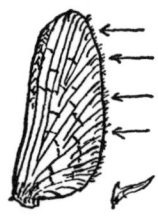

FIG. 14. Showing single intercalary veins and small narrow hindwings—with spur shaped costal projection.

Size-group 3. Medium-large. Large Spurwing.
            4. Medium to small. Small Spurwing.

Identification features
   Large Spurwing   wings blue-grey.
   Small Spurwing   wings very pale bluish grey.
   The males of both species have distinct orange-red eyes.

GROUP G. Duns with two wings only (no hindwings) and two tails.

Size-group 3. Medium-large to Medium. Pond Olive.
 4. Medium. Pale Evening dun.

Identification features
Pale Evening dun	body is a pale straw or pale olive, and the male has yellow eyes. Wings pale grey.
Pond Olive	body dark brownish olive; the male has dull orange-brown eyes, pale grey wings, and dark ringed tails.

KEY 1. THE FEMALE SPINNERS

SPINNERS WITH THREE TAILS

GROUP A. Spinners with upright hindwings (see Fig. 11) and three tails.

Size-group 1. Very large. Spent Gnat (Mayflies).
 3. Medium-large. Sherry spinner (B.W.O.), Claret spinner, Sepia spinner, Yellow Evening spinner, Turkey brown spinner.
 4. Medium. Purple spinner, Ditch spinner.

Identification features
Sherry spinner	body olive-brown to sherry red.
Yellow Evening spinner	body yellowish.
Ditch spinner	upright hindwings *with prominent costal projection*. (See Fig. 24)
Sepia spinner	body dark red-brown, smoky black area along front top of forewings. Upright hindwings *without costal projection*. (See Fig. 12)
Turkey Brown spinner	body dark brown. Upright hindwings *without costal projection*. (See Fig. 12).
Claret spinner	body brown with claret shade. Upright hindwings *without costal projection*. (See Fig. 12)
Purple spinner	body brownish purple. Upright hindwings *without costal projection*. (See Fig. 12)

IT IS DIFFICULT TO DIFFERENTIATE BETWEEN SOME OF THE ABOVE SPINNERS AS BODY COLORATION IS ONLY A VERY ROUGH GUIDE. IT IS THEREFORE ADVISABLE TO REFER TO THE NEXT CHAPTER FOR A POSITIVE CHECK.

GROUP B. Spinners with no hindwings and three tails.

Size-group 5. Small or very small. *Caënis*.

Identification features

Wings short and broad, body white. (The only spinner to retain the fine hairs along trailing edge of wings.)

SPINNERS WITH TWO TAILS—ALL REMAINING GROUPS

GROUP C. Spinners with upright hindwings (see Fig. 11) and two tails.

Size-group	2. Large.	Autumn spinner, Large Green spinner, March Brown spinner, Great Red spinner, Large Brook spinner.
	3. Medium-large.	Yellowstreak spinner, Yellow May spinner, Yellow Upright spinner.

Identification features

Dusky Yellowstreak spinner	body brown-olive. Yellow streak on thorax in front of each forewing root.
Large Brook spinner	body dark olive-brown diagonally banded red along sides. Mainly in small streams. Forewings have yellowish leading edge.
Large Green spinner	body olive-green with dark brown diagonal bands along sides. Also forewings are smoky-black along leading edge.
March Brown spinner	early season spinner, March to April. Body dark reddish brown, brown streak present in middle of each femur (top leg section).
Great Red spinner	body mahogany-red. Mid-season spinner.
Autumn spinner	body reddish brown. Late season spinner, August to September.
Yellow May spinner	body pale olive-yellow. Wings pale yellow along leading edge and eyes very pale blue.
Yellow Upright	mid-season spinner. Body yellow-olive to brown. Brown streak present in middle of each femur (top leg section).

GROUP D. Now included in above Group C.

IDENTIFICATION KEYS FOR UPWINGED FLIES

GROUP E. Spinners with double intercalary veins (paired), small oval-shaped hindwings and two tails. (See Fig. 13)

Size-group 3. Medium-large. Large Dark Olive spinner.
         4. Medium.   Medium Olive spinner. Dark Olive spinner.
         5. Small.   Pale Watery spinner. Small Dark Olive spinner. Iron Blue spinner.

Identification features

Large Dark Olive spinner	body red-brown. Tails ringed faint red-brown.
Dark Olive spinner	as above but darker and the hind wing lacks spur. (See Fig. 19)
Medium Olive spinner	body brown to red-brown. Wings have pale brown veining.
Pale Watery spinner	body pale golden brown to golden olive.
Small Dark Olive spinner	body medium red-brown, tinged olive. Wings have dark veining.
Iron Blue spinner	body dark claret-brown. Wings colourless.

The male spinners of all above have red-brown eyes, except the pale watery which has yellow eyes.

GROUP F. Spinners with single intercalary veins, very small narrow hindwings and two tails. (See Fig. 14)

Size-group 3. Medium-large.   Large Amber spinner.
         4. Medium to small. Little Amber spinner.

Identification features

Large Amber spinner	body rich amber colour ringed greyish.
Little Amber spinner	body pale yellow-amber colour, faintly ringed cream.

The male spinners of both these flies have distinct orange-red eyes.

GROUP G. Spinners with two wings only (no hindwings) and two tails.

Size-group 3. Medium large to medium. Pond Olive spinner.
         4. Medium.   Pale Evening spinner.

Identification features

Pond Olive spinner	Yellow-olive veins or shading along leading edge of wings. Body apricot coloured. (Male has orange-red eyes)
Pale Evening spinner	body pale golden olive-brown. (Male has yellow eyes)

It was decided not to include the male spinners in these KEYS TO SPINNERS, as this would have made the keys too complicated. In the majority of cases the male spinner is of little importance to the angler as they seldom fall spent on the water in any quantity.

There are, however, a few notable exceptions:

1. The male spinner of the Blue Winged Olive. This spinner has three tails, prominent red eyes, a dark red-brown body, and upright hindwings. Size, medium-large.
2. The male spinner of the Small Spurwing. This spinner has small narrow hindwings with a prominent spur (Fig. 33), a translucent white body with the last two or three segments orange-brown. Size medium to small.
3. The male spinner of the Large Spurwing. This spinner has a small narrow hindwing with prominent spur (Fig. 32), and a translucent white body with pale red ringing, and the last two or three segments dark amber. Size, medium-large.
4. The male spinner of the Iron Blue, known as the Jenny spinner. Has a translucent white body with the last three body segments dark orange-brown, and small oval hindwings with a spur on top edge. The thorax of this fly is shiny black. Size, small.

KEY NO. 2 (SIMPLIFIED)

The following visual keys based on colour and size have been included for the angler whose main interest is in actual fishing, and whose basic requirement is to be able to identify flies generally rather than by positive recognition of each individual species. It is only a rough guide due to the disparities in size and colour previously mentioned.

Many fly-fishermen are only interested in fly life as a means to an end, and quite rightly they wish to enjoy their sport without the detailed complication of entomology. In the past, too, many angling writers when writing articles or books on the subject of flies for the fisherman have been rather too technical, quoting scientific names and details that few anglers understand and some do not wish to understand.

This key is intended for anglers in this category.

As far as possible flies have been grouped together on the basis of colour and size. To quote an example, the Small Spurwing (*C. luteolum*) has been grouped with the Pale Evening dun. In general appearance it is so similar that without a detailed examination it is practically impossible to distinguish them. It is highly improbable that a trout would be able to spot the very small narrow hindwing on the former, or lack of same on the latter or the slight difference in veining of the main wings, which are the main points of positive identification. It will therefore be apparent that this key is not for the angler with a leaning towards a knowledge of entomology.

As this key is merely a rough guide to the identification of any given fly by the most obvious means, it is split into two simple parts: flies with two tails and flies with three tails. It is then narrowed down to groups similar in colour and size. Having established which group your fly comes into, according to size and basic coloration, you will find one or more flies mentioned in this group with variations on the basic colour, and from this the actual name of the fly can be established. Chapter XII can then be referred to, which will give you the suggested artificial pattern to use for the fly you have identified. Before attempting to use the key, the angler should refer to the methods at the beginning of these keys on how to distinguish a dun from a spinner (see page 36) and a male from a female (see page 26).

KEY 2. BASIC COLOUR KEY FOR THE UPWINGED FLIES

A. THE DUNS (MALE OR FEMALE)

1. Duns with Two Tails

Dark olive-green body. Size—large.
Duns in this group	Large Green dun.	Mottled wings and upright hindwings. Body colour dark olive-green, with diagonal brown bands along the sides.

Olive-brown body. Size—medium-large.
Duns in this group	Large Dark Olive.	Body colour medium to dark olive-brown.
	Olive Upright.	Body colour grey olive-brown (dark blue-grey wings and upright hindwings).
	Pond Olive.	Body colour dark brownish olive (no hindwings and size varies from medium-large to medium).

Olive-brown body. Size—medium.
Duns in this group	Medium Olive.	Body colour medium-olive.
	Dark Olive.	Body colour dark olive-brown.

Pale olive-grey body. Size—medium-large.
Duns in this group	Large Spurwing.	Body colour pale olive-grey (wings blue-grey).

Pale watery olive body. Size—medium.

Duns in this group	Small Spurwing.	Body colour pale olive (size medium to small).
	Pale Evening dun.	Body colour pale straw or pale olive (no hindwings and male has yellow eyes).

Pale watery grey-olive body. Size—small.

Duns in this group	Small Dark Olive.	Body colour grey-olive.
	Pale Watery.	Body colour pale grey-olive (male has yellow eyes).

Dark grey body. Size—medium-large.

Duns in this group	Dusky Yellowstreak.	Body colour dark greyish brown (dark grey wings and upright hindwings). Note yellow streak in front of each wing root on body.

Dark body. Size—small.

Duns in this group	Iron Blue.	Body colour very dark brown-olive (grey-blue wings).

Brownish body. Size—large.
All duns in this group have mottled wings and upright hindwings.

	March Brown.	Body colour dull brown. (A spring fly.)
	Autumn dun.	Body colour pale olive-brown. (An autumn fly.)
	Large Brook dun.	Body colour drab olive-brown.
	Late March Brown.	Body colour dull brown. (A mid-season fly.)

Yellow body. Size—medium-large.

Duns in this group	Yellow May dun.	Colour pale yellow body and wings, with upright hindwings.

II. Duns with Three Tails

All have upright hindwings except *Caënis* which has none.

Olive-green body. Size—medium-large.

Duns in this group	Blue Winged Olive.	Body colour olive-green to rusty brown olive (wings dark blue-grey). The male fly is much darker (and browner) in the body.

Dark brown body. —Size medium-large.

Duns in this group	Sepia dun.	Body colour dark sepia brown (heavily veined wings).
	Turkey Brown.	Body colour dark blackish brown (mottled wings).
	Claret dun.	Body colour dark brown (dark grey wings).

IDENTIFICATION KEYS FOR UPWINGED FLIES 47

Dark brown-purple body. Size—medium.
Duns in   Purple Dun.   Body colour dark brown-purple
this group   (blackish grey wings).

Yellow body. Size—medium-large.
Duns in   Yellow Evening dun.   Body colour pale yellow (yellow
this group   wings).

Cream body. Size—very large.
Duns in   Mayflies.   Body colour creamy yellow.
this group

Cream body. Size—small or very small.
Duns in   *Caënis.*   Body colour dull cream (no hind-
this group   wings).

B. THE FEMALE SPINNERS

I. Spinners with Two Tails

Mahogany-red body. Size—large, with upright hindwings.
Spinners in   Great Red spinner.   Body colour mahogany-red.
this group

Red-brown body. Size—large, with upright hindwings.
Spinners in   Autumn spinner.   Body colour red-brown.
this group   (Late season spinner.)
           Large Brook spinner.   Body colour red-brown, diagonal bands.
           March Brown spinner.   Body colour dark red-brown. (Early season spinner.)

Red-brown body. Size—medium-large.
Spinners in   Large Dark Olive spinner.   Body colour red-brown.
this group

Red-brown body. Size—medium.
Spinners in   Medium Olive spinner.   Body colour brown to red-brown.
this group
           Dark Olive spinner.   Body colour dark red-brown

Red-brown body. Size—small or very small.
Spinners in   Small Dark Olive spinner.   Body colour reddish brown.
this group   Iron Blue spinner.   Body colour dark claret-brown.

Amber body ringed cream. Size—medium-large.
Spinners in   Large Amber spinner.   Body colour rich amber with
this group   greyish ringing.

Yellow-amber body. Size—medium to small.
Spinners in   Little Amber spinner.   Body colour pale yellow-
this group   amber, faintly ringed cream.

Apricot body. Size—medium to medium-large.
Spinners in Pond Olive spinner. Body colour apricot (no hindwings). (Occurs in still and slow-flowing waters.)
this group

Golden olive body. Size—medium.
Spinners in Pale Evening spinner. Body colour palest golden olive brown (no hindwings).
this group

Golden body. Size—small.
Spinners in Pale Watery spinner. Body colour golden olive.
this group

Olive-green body. Size—large, with upright hindwings.
Spinners in Large Green spinner. Body colour olive-green with diagonal brown bands along sides.
this group

Brown-olive body. Size—medium-large, with upright hindwings.
Spinners in Dusky Yellowstreak spinner. Body colour brown-olive. Note distinct yellow streak in front of wing root on each side of body.
this group

Pale yellow body. Size—medium-large, with upright hindwings.
Spinners in Yellow May spinner. Body colour pale olive-yellow, wings pale yellow along leading edge.
this group

Yellow-olive body. Size—medium-large, with upright hindwings.
Spinners in Yellow Upright spinner. Body colour yellow-olive to brown.
this group

II. Spinners with Three Tails

All have upright hindwings except *Caënis* which has none.

Sherry-red body. Size—medium-large.
Spinners in Sherry spinner. Body colour sherry to lobster-red.
this group

Brown body. Size—medium-large.
Spinners in Sepia spinner. Body colour dark red-brown.
this group Claret spinner. Body colour brown with claret shade.

Turkey Brown spinner. Body colour dark brown.

Brownish purple body. Size—medium.
Spinners in Purple spinner. Body colour brownish purple.
this group

Yellow body. Size—medium-large.
Spinners in Yellow Evening spinner. Body colour yellowish.
this group

Cream body. Size—very large. Spinners in this group	*Mayfly*.	Body colour cream.
Cream body. Size—small or very small. Spinners in this group	*Caënis*.	Body colour white (no hindwings). (This is the only spinner to retain the fine hairs along trailing edge of wings.)

Some of the features mentioned in this chapter are tabulated below in a simple form which should assist in a quick identification of some species of both male and female dun and spinner (except where otherwise stated).

Constant feature	*Species*
Yellow streak each side of thorax in front of forewing root	*H. lateralis*
Brown streak in middle of each top leg section (femur)	*R. haarupi* *R. semicolorata*
Mottled wings with clear area in centre	*R. haarupi* *E. torrentis* } duns *E. insignis* } only *P. submarginata*
Yellowish leading edge to wings	*E. torrentis* *C. dipterum*—female spinner only *E. notata*—male and female spinner only *H. sulphurea*—female spinner only
Blackish leading edge to wings	*L. marginata*—spinners only *E. insignis*—spinners only *H. sulphurea*—male spinner only
Upright hindwings without costal projection	*L. marginata* *P. submarginata* *L. vespertina* *P. cincta*
Distinct yellow eyes	*P. pseudorufulum* } males *B. bioculatus* } only
Two parallel red lines transversing ventral segments	*C. dipterum*—females only

D

CHAPTER IV

DETAILED DESCRIPTIONS OF THE UPWINGED FLIES (EPHEMEROPTERA)

FOLLOWED BY WING ILLUSTRATIONS SHOWING VEINING

The following detailed descriptions of the Upwinged flies have been provided to assist in positive identification. It is emphasized that the colours given are as seen on the majority of specimens that I have myself examined. Apart from the fact that colours of these flies vary greatly, many of us describe colours as we see them in different ways, and my interpretation of these colours may not be in accordance with other authorities. Most species of Upwinged flies of interest to anglers have been included in these descriptions, and at the end of this chapter are sketches and photographs showing the veining on the wings of many of them. These can often be of great assistance where identification is uncertain. It should be noted that generally the female is larger than the male.

In the following descriptions the body of the fly is in many cases divided into two parts—the DORSUM (top) and the VENTER (underpart). Where it is not divided the coloration is overall.

In many cases the rear edge of the body (abdomen) segments are of a different colour from the major portion of the body segments. For simplicity, in the detailed descriptions these lighter or darker areas round the body where particularly noticeable are referred to as "ringing" or "joinings".

The hindwings when present can often be of considerable help in identification. Many of these have a costal projection along the top or leading edge, and on those hindwings where it is very distinct it is referred to as a spur.

N.B. As previously explained, the true colour of the eyes of the male duns may not be apparent when freshly hatched. It is therefore advisable to let a short period elapse before examining freshly hatched males.

1. IRON BLUE DUN—FEMALE
   Baëtis niger

2. IRON BLUE DUN—MALE
   Baëtis niger

3. MEDIUM OLIVE DUN—FEMALE
   Baëtis vernus

4. PALE WATERY DUN—FEMALE
   Baëtis bioculatus

5. PALE WATERY DUN—MALE
   Baëtis bioculatus

6. MEDIUM OLIVE DUN—MALE
   Baëtis vernus

7. SMALL DARK OLIVE DUN—FEMALE
   Baëtis scambus

8. SMALL DARK OLIVE DUN—MALE
   Baëtis scambus

9. PALE EVENING DUN—FEMALE
   Procloëon pseudorufulum

10. DARK OLIVE DUN—FEMALE
    Baëtis atrebatinus

11. POND OLIVE DUN—MALE
    Cloëon dipterum

12. GRAVEL BED FEMALE
    Hexatoma fuscipennis

13. NYMPH—AGILE DARTER
    Cloëon dipterum

14. COW DUNG FLY
    Cordilura Spp

15. LARVA OF A WATER BEETLE

PLATE I.   DUNS, ETC. x 1·6

DESCRIPTIONS OF UPWINGED FLIES

Ephemera danica (THREE-TAILS)

DUN

Mayfly, male

EYES: Very dark brown.
WINGS: Heavily veined brown on a grey (tinged yellow) ground, with dark markings.
BODY: Dorsum—greyish white with brown blotches.
Venter—greyish white with constant brown marks.
LEGS: Dark brown.
TAILS: Dark grey to black.
SIZE: Very large.

Mayfly, female
(Plate IV. 41)

EYES: Black-brown.
WINGS: Grey tinged blue-green with dark markings. Heavily veined blackish. Yellow tinged along leading edge.
BODY: Dorsum—yellowish cream with constant brown marks. (See Fig. 46.)
Venter—slightly paler with brown markings. (See Plate X. 140.)
LEGS: Creamy olive with black tinges.
TAILS: Very dark grey to black.
SIZE: Very large.
Remarks: The thorax is a distinct pale orange colour, and they have large upright hindwings with a costal projection. (Fig. 11)

SPINNER

Spent Gnat, male

EYES: Black-brown.
WINGS: Transparent with brownish tint. Heavily veined brown with several almost black patches.
BODY: Dorsum—creamy white, last three segments brownish.
Venter—Cream with constant brown marks and grey ringing, last three segments brown.
LEGS: Forelegs black-brown, remainder dark olive brown.
TAILS: Dark brown.
SIZE: Very large.

Spent Gnat, female

EYES: Black-brown.
WINGS: Transparent but with a bluish tint. Veined brown with several almost black patches.

BODY: Dorsum—pale cream. Three tail segments with brown streaks.
Venter—cream with constant brown markings.
LEGS: Dark olive-brown.
TAILS: Dark brown.
SIZE: Very large.
Remarks: The thorax of these spinners is sooty-black, and the forelegs are very long.

Baëtis rhodani (TWO-TAILS)

DUN

Large Dark Olive, male
(Plate II. 17)

EYES: Dull brick-red.
WINGS: Pale grey with pale brown veins.
BODY: Dorsum—dark olive-brown or dark olive-green.
Venter—dark olive-grey. Last segment yellow-olive.
LEGS: Forelegs grey-olive. Remainder pale olive. Feet dull black.
TAILS: Dull grey.
SIZE: Medium large.
Remarks: The body colour of these flies is considerably lighter in the warmer weather in late spring and again when they reappear in early autumn.

Large Dark Olive, female
(Plate II. 16)

EYES: Greenish.
WINGS: Pale grey with pale brown veins.
BODY: Dorsum—dark olive-brown or dark olive-green.
Venter—dark olive or olive.
LEGS: All legs pale green-olive. Feet dull black.
TAILS: Dull grey.
SIZE: Medium-large.
Remarks: The female is usually larger than the male. Both dun and spinner have double marginal intercalary veins and small oval-shaped hind-wings with a distinct spur on the top edge near wing root (see Fig. 13). The side of the thorax of this fly is a pale orange colour.

SPINNER

Large Dark Olive spinner, male
(Plate VII. 82)

EYES: Very dark red-brown.
WINGS: Transparent with brown longitudinal veins.

DESCRIPTIONS OF UPWINGED FLIES 53

BODY:	Dorsum—pale olive tinged brown. Last three segments orange-brown.
	Venter—pale olive. Last three segments orange-brown.
LEGS:	Forelegs dark grey-olive. Remainder pale olive. Feet grey-black.
TAILS:	Pale grey ringed red-brown.
SIZE:	Medium-large.

Large Red spinner, female
(Plate VII. 81)

EYES:	Black-brown.
WINGS:	Transparent with brown longitudinal veins.
BODY:	Dorsum—red-brown with paler joinings. Venter—pale olive.
LEGS:	Forelegs dark olive. Remainder brown-olive. Feet dark grey.
TAILS:	Olive-grey ringed red-brown.
SIZE:	Medium-large.

Remarks: These are the largest of the olive spinners and the faint ringing of the tails is a help in identification.

Baëtis pumilus or niger (TWO-TAILS)

DUN

Iron Blue, male
(Plate I. 2)

EYES:	Dull red-brown.
WINGS:	Dull grey-blue.
BODY:	Grey-black tinged with dark olive or dark brown.
LEGS:	Dark olive-brown.
TAILS:	Dark olive-grey.
SIZE:	Small.

Remarks: A small to very small brown-black fly.

Iron Blue, female
(Plate I. 1)

EYES:	Dull yellow-green.
WINGS:	Dull grey-blue.
BODY:	Very dark brown olive.
LEGS:	Pale olive-brown.
TAILS:	Dark grey.
SIZE:	Small.

Remarks: A small to very small fly. Not quite as dark as the male.

Both dun and spinner have double marginal intercalary veins and small oval-shaped hindwings with a distinct spur on the top edge near the wing root. (See Fig. 13)

SPINNER

Iron Blue,   EYES: Very dark red-brown.
  (Jenny) spinner, WINGS: Transparent.
  male   BODY: Translucent white. Last three segments dark orange-brown.
(Plate VII. 86)
  LEGS: Pale grey, forelegs darker.
  TAILS: White-grey.
  SIZE: Small.
  Remarks: The top of the thorax of this male spinner is shiny black.

Little Claret   EYES: Black.
  spinner, female WINGS: Transparent.
(Plate VII. 83)   BODY: Dark claret-brown considerably paler underneath.
  LEGS: Pale grey-olive.
  TAILS: Pale grey.
  SIZE: Small.

Baëtis vernus (TWO-TAILS)

DUN

Medium Olive,   EYES: Dull red-brown.
  male   WINGS: Dull grey with tinges of golden olive along the veins.
(Plate I. 6)
  BODY: Dorsum—dark grey-olive to medium olive. Venter—pale grey-olive. Last segment yellow-olive.
  LEGS: Forelegs dark olive, remainder pale olive. Feet brown-black.
  TAILS: Grey.
  SIZE: Medium.

Medium Olive,   EYES: Dull yellow-green.
  female   WINGS: Dull grey with light brown veining.
(Plate I. 3)   BODY: Dorsum—brown-olive to medium olive. Venter—pale yellow-olive.
  LEGS: Pale brown-olive. Feet brown-black.
  TAILS: Grey.
  SIZE: Medium.
  Remarks: Both dun and spinner have double marginal intercalary veins and small oval-shaped hindwings with a distinct spur on the top edge near wing root. (See Fig. 13)

SPINNER

Medium Olive
spinner,
male
(Plate VII. 85)

EYES: Red-brown.
WINGS: Transparent with light brown veins.
BODY: In some specimens the whole of the body is a dark red-brown, in others only the last three segments, with the middle body a grey-olive colour.
LEGS: Grey-olive.
TAILS: Off-white.
SIZE: Medium.
Remarks: Similar to the Large Olive spinner but noticeably darker in the body, and tails almost water white.

Red spinner,
female
(Plate VII. 84)

EYES: Brown.
WINGS: Transparent with palest brown veins.
BODY: Dorsum—Varies from a yellow-brown to red-brown.
Venter—Paler with last two segments yellowish.
LEGS: Grey-olive.
TAILS: Off-white.
SIZE: Medium.

Baëtis scambus (TWO-TAILS)

DUN

Small Dark Olive,
male
(Plate I. 8)

EYES: Dull orange-red.
WINGS: Medium to dark grey.
BODY: Dorsum—grey-olive.
Venter—last two segments yellowish.
LEGS: Pale yellow-olive. Feet black.
TAILS: Grey.
SIZE: Small to very small.
Remarks: The smallest of the olives. Some of the males are as small as the larger species of *Caënis*.

Small Dark Olive,
female
(Plate I. 7)

EYES: Black.
WINGS: Medium to dark grey.
BODY: Dorsum—grey-olive.
Venter—paler with last two segments yellowish.
LEGS: Pale yellow-olive, black feet.
TAILS: Grey.
SIZE: Small.

Remarks: Both dun and spinner have double marginal intercalary veins and small oval-shaped hindwings with a distinct spur on the top edge near wing root. (See Fig. 13)

SPINNER

Small Dark Olive spinner, male
(Plate VIII. 103)

EYES: Orange-red.
WINGS: Transparent.
BODY: Translucent cream, last three segments opaque orange-brown.
LEGS: Olive-brown.
TAILS: Greyish white.
SIZE: Small to very small.
Remarks: Similar to the male (Jenny) spinners of the Iron Blue or the Small Spurwing, but the thorax is a very dark brown or brown-black.

Small Red spinner, female
(Plate VIII. 102)

EYES: Black.
WINGS: Transparent with darkish veins.
BODY: Varies from dark brown tinged with olive to deep red-brown, last two segments yellowish.
LEGS: Olive-brown.
TAILS: Greyish white.
SIZE: Small.
Remarks: The body colour of this female spinner varies a great deal. Specimens have been observed with only a little red or brown in the body; very dark olive being the predominant colour.

Baëtis bioculatus (TWO-TAILS)

DUN

Pale Watery, male
(Plate I. 5)

EYES: Lemon-yellow to orange-yellow.
WINGS: Pale grey, veins pale brown.
BODY: Pale grey-olive, last two segments pale yellow.
LEGS: Pale olive. Feet dark grey.
TAILS: Grey.
SIZE: Small.
Remarks: A rather distinctive fly. The greyish body contrasts rather sharply with the two pale yellow segments at the tail and of course the two yellow eyes are very conspicuous.

Pale Watery, female
(Plate I. 4)

EYES: Dull yellow-green.
WINGS: Pale grey, veins very pale brown.
BODY: Dorsum—pale grey-olive.
Venter—pale watery-olive, last two segments pale yellow-olive.
LEGS: Pale olive. Feet dark grey.
TAILS: Grey.
SIZE: Small.
Remarks: Very similar to the Small Dark Olive. In fact it is extremely difficult to tell these two female flies apart. However, the Small Dark Olive is usually a little darker and smaller.

Both dun and spinner have double marginal intercalary veins and small oval-shaped hindwings, with a distinct spur on the top edge near wing root. (See Fig. 13)

SPINNER

Pale Watery spinner, male
(Plate VII. 88)

EYES: Lemon-yellow to orange-yellow.
WINGS: Transparent.
BODY: Creamy translucent white tinged olive. Last three segments orange-brown.
LEGS: Pale white-olive.
TAILS: Greyish white.
SIZE: Small.
Remarks: This spinner is very similar to the Small Dark Olive, and the Iron Blue male (Jenny) spinners. However, it is easily distinguished by its yellowish eyes.

Golden spinner, female
(Plate VII. 87)

EYES: Brown-black.
WINGS: Transparent.
BODY: Pale golden brown to golden olive, with the last three segments a slightly darker shade.
LEGS: Pale olive-brown.
TAILS: Greyish white.
SIZE: Small.
Remarks: The body coloration of this spinner varies from an almost golden yellow to a pale olive-brown.

Baëtis atrebatinus (TWO-TAILS)

DUN

Dark Olive, male

EYES: Pale red-brown.
WINGS: Grey.
BODY: Dorsum—dark olive-brown.
Venter—dark olive. Last segment yellowish.
LEGS: Pale olive with brown feet.
TAILS: Dark grey-olive.
SIZE: Medium.
Remarks: Very similar to the Large Dark Olive, but of an overall darker appearance, and a little smaller.

Dark Olive, female
(Plate I. 10)

EYES: Dark greenish olive.
WINGS: Grey.
BODY: Dorsum—dark olive-brown.
Venter—dark yellow-olive. Last segment yellowish.
LEGS: Pale olive with brown feet.
TAILS: Grey.
SIZE: Medium.
Remarks: Both dun and spinner have double marginal intercalary veins with oval shaped hindwings, but without costal projection (spur) on top edge. (See Figs. 13 and 19)

SPINNER

Dark Olive, male spinner

EYES: Red-brown.
WINGS: Transparent, veins tinged pale brown.
BODY: Very pale grey-green with last three segments brownish amber.
LEGS: Olive to grey.
TAILS: Grey-white faintly ringed red.
SIZE: Medium.

Dark Olive, female spinner

EYES: Black-brown.
WINGS: Transparent, veins faintly ringed pale brown.
BODY: Dorsum—dark red-brown with paler ringing.
Venter—pale olive.
LEGS: Brown-olive to grey.
TAILS: Olive-grey faintly ringed red.

SIZE: Medium.
Remarks: Very similar to Large Red spinner, but the tails are not so heavily ringed and the veining of the wings is much more subdued.

Ephemerella notata (THREE-TAILS)

DUN

Yellow Evening dun, male

- EYES: Amber-orange.
- WINGS: Pale grey with yellowish veining. Viewed from a distance the wings appear pale yellow.
- BODY: Dorsum—pale yellow with last three segments light amber.
  Venter—as above but paler.
- LEGS: Yellowish.
- TAILS: Yellowish, ringed brown.
- SIZE: Medium-large.

Yellow Evening dun, female

- EYES: Pale green.
- WINGS: Palest yellow-grey with yellowish veining.
- BODY: Yellow.
- LEGS: Pale yellow to greyish.
- TAILS: Yellow, ringed brown.
- SIZE: Medium-large.

Remarks: These duns are similar in appearance but smaller than the Yellow May dun, and they have three tails. On the venter of each body segment, both duns and spinners have a clearly defined pattern of dots and dashes. (See Fig. 15) Both duns and spinners have upright hindwings *with costal projection*. (See Fig. 11)

SPINNER

Yellow Evening spinner, male

- EYES: Medium orange.
- WINGS: Transparent with veins along leading edge yellowish.
- BODY: Dorsum—yellow-olive tinged grey. Last three segments brown-olive.
  Venter—as above but paler.
- LEGS: Olive-yellow.
- TAILS: Yellow, ringed red-brown.
- SIZE: Medium-large.

Yellow Evening   EYES: Greenish.
spinner, female   WINGS: Transparent with veins along leading edge yellowish.
     BODY: Dorsum—yellow-olive. Last three segments brown-olive.
            Venter—as above but paler.
     LEGS: Olive-yellow.
     TAILS: Yellow, ringed red-brown.
     SIZE: Medium-large.

Ephemerella ignita (THREE-TAILS)

DUN

Blue Winged Olive,   EYES: Red.
male   WINGS: Dark blue-grey.
(Plate II. 23)   BODY: Orange-brown, varies to an almost olive-brown in some specimens, last segment yellowish.
     LEGS: Brown-olive. Feet dark grey.
     TAILS: Dark grey, ringed brownish.
     SIZE: Medium-large.
     Remarks: A most distinctive fly. Brownish body with three tails and dark blue-grey wings. Once correctly identified, not likely to be confused with any other fly.

Blue Winged Olive,   EYES: Dark greenish-black.
female   WINGS: Dark blue-grey.
(Plate II. 22)   BODY: Early in June the body is a bright green-olive, but through the season it darkens to a dull grey-olive and late in the year to a rusty brown-olive.
     LEGS: Dark olive. Feet grey.
     TAILS: Pale grey-brown, ringed dark brown.
     SIZE: Medium-large.
     Remarks: Both dun and spinner have upright hindwings with costal projection. (Fig. 11) The forewings are also large for the size of the fly and are backward raked.

SPINNER

Blue Winged Olive   EYES: Bright red.
spinner male   WINGS: Transparent with light brown veins.
(Plate VII. 94)   BODY: Dark brown to a rich red-brown on some specimens.

16. LARGE DARK OLIVE DUN-FEMALE *Baëtis rhodani*

17. LARGE DARK OLIVE DUN-MALE *Baëtis rhodani*

18. LARGE SPURWING DUN-FEMALE *Centroptilum pennulatum*

19. SMALL SPURWING DUN-FEMALE *Centroptilum luteolum*

20. SMALL SPURWING DUN-MALE *Centroptilum luteolum*

21. LARGE SPURWING DUN-MALE *Centroptilum pennulatum*

22. BLUE-WINGED OLIVE DUN-FEMALE *Ephemerella ignita*

23. BLUE-WINGED OLIVE DUN-MALE *Ephemerella ignita*

24. CLARET DUN-FEMALE *Leptophlebia vespertina*

25. OLIVE UPRIGHT DUN-FEMALE *Rhithrogena semicolorata*

26. OLIVE UPRIGHT DUN-MALE *Rhithrogena semicolorata*

27. DUSKY YELLOWSTREAK SPINNER FEMALE *Heptagenia lateralis*

28. POND OLIVE DUN-FEMALE *Cloëon dipterum*

29. BLACK GNAT *Bibio johannis*

30. FEMALE BLUE-WINGED OLIVE WITH EGG BALL

PLATE II. DUNS AND SPINNERS, ETC. × 1·6

DESCRIPTIONS OF UPWINGED FLIES 61

LEGS:	Pale brown.
TAILS:	Pale fawn with black rings.
SIZE:	Medium-large.

Sherry spinner, female
(Plate VII. 93)

EYES:	Greenish brown.
WINGS:	Transparent with pale brown veins.
BODY:	Varies in different specimens from an olive-brown to an almost lobster-red or sherry-red.
LEGS:	Pale brown.
TAILS:	Olive-grey faintly ringed brown.
SIZE:	Medium-large.

Remarks: The body colour of this spinner varies tremendously, even among individual flies in the same swarm.

Centroptilum luteolum (TWO-TAILS)

DUN

Small Spurwing, male
(Plate II. 20)

EYES:	Orange-red.
WINGS:	Very pale grey, sometimes tinged blue-white. Veins pale olive.
BODY:	Dorsum—pale olive-grey. Venter—olive-brown, last two segments orange-brown.
LEGS:	Brown-olive. Feet smoky grey.
TAILS:	Grey.
SIZE:	Medium to small.

Small Spurwing, female
(Plate II. 19)

EYES:	Pale green.
WINGS:	Very pale grey, sometimes tinged blue-white. Veins pale olive.
BODY:	Dorsum—pale brown-olive to pale watery olive. Venter—pale olive.
LEGS:	Very pale olive. Feet grey.
TAILS:	Grey.
SIZE:	Medium to small.

Remarks: The flies, which appear in the early summer, are usually darker than the pale watery-coloured specimens which are common during the late summer. The eyes of the male are normally a brighter red than any of the other olives.

Both dun and spinner have single marginal intercalary veins and very small narrow hindwings with prominent spur. (See Fig. 25)

SPINNER

Small Spurwing spinner, male
(Plate VII. 92)

EYES: Bright orange-red.
WINGS: Transparent veins faintly olive.
BODY: Translucent watery white, last three segments pale orange-brown.
LEGS: Very pale olive.
TAILS: Pale grey-white.
SIZE: Medium to small.
Remarks: This male spinner has a very pale orange-brown thorax and is similar to the pale watery male spinner, but can be easily distinguished by its bright reddish eyes (the pale watery has yellow eyes).

Little Amber spinner, female
(Plate VII. 89)

EYES: Brown-black.
WINGS: Transparent veins faintly olive.
BODY: Dorsum—varies from a pale yellow-brown to a pale amber colour when fully spent. Faintly ringed cream.
Venter—is a creamy yellow with last two segments light amber.
LEGS: Pale olive-brown.
TAILS: Pale olive-white.
SIZE: Medium to small.
Remarks: The body colour of this spinner is quite distinctive.

Centroptilum pennulatum (TWO-TAILS)

DUN

Large Spurwing, male
(Plate II. 21)

EYES: Dull orange.
WINGS: Blue-grey.
BODY: Dorsum—pale olive-brown or grey with last three segments amber coloured.
Venter—grey-white with last three segments straw colour.
LEGS: Pale olive-brown. Feet grey.
TAILS: Grey.
SIZE: Medium-large.

Large Spurwing, female
(Plate II. 18)

EYES: Yellow-green.
WINGS: Blue-grey.
BODY: Dorsum—pale olive-grey to creamy grey.
Venter—pale straw.

LEGS:	Upper legs olive, lower grey-white. Feet grey.
TAILS:	Grey.
SIZE:	Medium-large.

Remarks: A very distinctive fly easily recognized by its almost white body and dark blue-grey wings.

Both dun and spinner have single marginal intercalary veins and very small narrow hindwing with prominent spur. (See Fig. 26)

SPINNER

Large Spurwing, male spinner
(Plate VII. 91)

EYES:	Bright Orange.
WINGS:	Transparent.
BODY:	Dorsum—translucent white with pale red ringing, last three segments dark amber.
	Venter—silver-white.
LEGS:	Pale grey.
TAILS:	Grey-white.
SIZE:	Medium-large.

Remarks: The thorax of this spinner is a pale brown or fawn colour.

Large Amber spinner
(Plate VII. 90)

EYES:	Yellow-grey going blackish with age.
WINGS:	Transparent, veins palest olive.
BODY:	Dorsum—varies from a pale amber-flecked olive to a deep rich amber with greyish ringing.
	Venter—whitish olive.
LEGS:	Olive-grey.
TAILS:	Pale grey.
SIZE:	Medium-large.

Remarks: A large amber-coloured spinner with transparent wings.

Procloëon pseudorufulum (TWO-TAILS)

DUN

Pale Evening dun, male

EYES:	Dull yellow.
WINGS:	Pale grey.
BODY:	Dorsum—straw-coloured, often tinged with pale brown.
	Venter—very pale straw.
LEGS:	Pale olive, becoming greyish towards the feet.
TAILS:	Olive-grey.
SIZE:	Medium.

Pale Evening dun, EYES: Olive-green.
female WINGS: Pale grey.
(Plate I. 9) BODY: Dorsum—pale straw or olive tinged with red-brown patches.
Venter—very pale straw.
LEGS: Pale olive, becoming greyish towards the feet.
TAILS: Olive-grey.
SIZE: Medium.
Remarks: A very pale-coloured fly, normally only hatches late in the day.
Both dun and spinner have single marginal intercalary veins and no hindwings.

SPINNER

Pale Evening EYES: Pale yellow.
spinner, male WINGS: Transparent.
BODY: Dorsum—translucent grey-white with last three segments amber to orange.
Venter—very pale grey.
LEGS: Pale whitish olive.
TAILS: Grey-white.
SIZE: Medium.
Remarks: This male spinner is similar to the Pale Watery spinner, except that the thorax and last three segments of the body are of a lighter orange-brown. Also the body segments are sometimes tinged pale red-brown.

Pale Evening EYES: Dark yellow-olive.
spinner, female WINGS: Transparent.
(Plate VIII. 100) BODY: Dorsum—palest golden olive-brown.
Venter—very pale translucent grey-white.
LEGS: Pale whitish olive.
TAILS: Grey-white.
SIZE: Medium.
Remarks: Both male and female dun and spinner often have a bright green stain along the leading edge of the wings which have six to eight small cross veins along top leading edge. (Fig. 34)

Cloëon dipterum (TWO-TAILS)

DUN

Pond Olive, male EYES: Dull orange-brown, with two faint red lines across centre.
(Plate I. 11)

WINGS: Pale grey.
BODY: Dorsum—dull grey-olive, last three segments dull grey-brown.
Venter—dull grey-olive, last segment yellow.
LEGS: Pale watery-white. Faint reddish marks top of forelegs.
TAILS: Pale grey ringed brown.
SIZE: Medium.

Pond Olive female
(Plate II. 28)

EYES: Dull green with two faint red lines across centre.
WINGS: Darkish grey.
BODY: Dorsum—dark brownish olive, streaked red in some cases.
Venter—green with a brown tinge.
LEGS: Pale watery olive, feet grey. Top of forelegs suffused reddish.
TAILS: Pale grey ringed black-brown.
SIZE: Medium to medium-large.
Remarks: Both dun and spinner have single marginal intercalary veins and are without hindwings. They have three to five small cross veins along top leading edge (see Fig. 37) of forewings. Also the female dun and spinner often have two parallel red lines which transverse the length of the ventral segments (see Plate VIII. 110).

Pond Olive spinner, male
(Plate VIII. 104)

EYES: Orange-red, with two faint red lines across centre.
WINGS: Transparent, with main veins along leading edge pale brown.
BODY: Dorsum—dull translucent cream, last three segments dark brown.
Venter—dull greyish cream.
LEGS: Pale grey-white.
TAILS: Grey-white ringed brown.
SIZE: Medium.

Apricot spinner, female
(Plate VI. 79)

EYES: Black.
WINGS: Transparent with red-brown veining. The leading edge of each wing is yellow-olive tinged red.
BODY: Dorsum—varies from an apricot colour streaked red to red-brown, tinged dark yellow.

Venter—olive.
LEGS: Forelegs olive-ringed red, remainder bright olive-green.
TAILS: Very pale grey ringed brown.
SIZE: Medium to medium-large.
Remarks: A very distinctive spinner with its ringed tail and red blotches on various parts of its body. Further unusual features are the broad bands of yellow along the leading edge of the wings and the yellow thorax.

Rhithrogena haarupi (TWO-TAILS)

DUN

March Brown, male
(Plate III. 34)

EYES: Very dark green with black bar across centre.
WINGS: Membrane pale fawn with heavy dark brown veins, giving the wing a mottled appearance.
BODY: Dorsum—dark brown ringed straw colour.
Venter—similar but a slightly paler shade of brown.
LEGS: Pale brown with feet brown to black.
TAILS: Dark brownish grey.
SIZE: Large.

March Brown, female
(Plate III. 33)

EYES: Very dark green with a black bar across centre.
WINGS: Membrane fawn with heavy dark brown veins giving the wings a mottled appearance.
BODY: Dorsum—dull brown ringed straw colour.
Venter—paler but tinged reddish brown.
LEGS: Pale olive-brown with feet brown to black. The forelegs are darker.
TAILS: Dark brown.
SIZE: Large.
Remarks: A useful point of identification on the duns are two clear patches in the centre of the mainwings devoid of cross veins. Both duns and spinners have upright hindwings with a costal projection (Fig. 11).

DESCRIPTIONS OF UPWINGED FLIES

SPINNER

March Brown, male spinner
- EYES: Black.
- WINGS: Transparent with brown veining.
- BODY: Dorsum—dark brown ringed cream. Venter—slightly paler.
- LEGS: Forelegs brown-black, others pale brown.
- TAILS: Brown.
- SIZE: Large.

March Brown, female spinner
(Plate VI. 76)
- EYES: Dull green with dark brown bar across centre.
- WINGS: Transparent with dark brown veining.
- BODY: Dorsum—dark reddish brown ringed straw. Venter—slightly paler.
- LEGS: Tops brown-olive fading to a pale yellow-olive at feet.
- TAILS: Chestnut-brown.
- SIZE: Large.

Remarks: Both duns and spinners have a distinct brown streak in the middle of each top leg section (femur).

Rhithrogena semicolorata (TWO-TAILS)

DUN

Olive Upright, male
(Plate II. 26)

- EYES: Dark black-olive.
- WINGS: Dark blue-grey.
- BODY: Dorsum—dark grey-olive, ringed olive. Venter—pale grey-olive.
- LEGS: Pale olive-brown.
- TAILS: Dark grey-brown.
- SIZE: Medium-large.

Olive Upright, female
(Plate II. 25)
- EYES: Dull green with a brown bar across centre.
- WINGS: Dark blue-grey, trailing edge of hindwing buff colour.
- BODY: Dorsum—grey olive-brown tinged reddish, ringed olive. Venter—very pale yellow-olive.
- LEGS: Pale olive-brown.
- TAILS: Grey-brown.
- SIZE: Medium-large.

Remarks: Both male and female duns have a distinct

brown streak in the middle of each top leg section (femora), and the side of the thorax is of a distinct orange colour. They have upright hindwings, with a costal projection. (Fig. 31)

SPINNER

Yellow Upright, male spinner
(Plate VIII. 99)

EYES: Olive-yellow with dark bar across base.
WINGS: Transparent veined brown. Lower half of forewings are sometimes a smoky bronze colour.
BODY: Dorsum—brown-olive, ringed olive.
Venter—grey-olive with an overall yellow tinge.
LEGS: Forelegs grey-olive, others olive.
TAILS: Pale grey-buff.
SIZE: Medium-large.
Remarks: The male spinner in flight often appears to be of a sulphury yellow colour. In general appearance they are similar to the male Yellow May spinner.

Yellow Upright, female spinner
(Plate VI. 75)

EYES: Dull olive-yellow with dark bar across centre.
WINGS: Transparent veined pale brown. Lower half of forewings are sometimes a smoky bronze colour.
BODY: Dorsum—dull yellow-olive ringed brown-olive.
Venter—cream-olive.
LEGS: Pale olive.
TAILS: Pale buff ringed very faintly red-brown.
SIZE: Medium-large.
Remarks: The hindwings of these spinners are very lightly veined and difficult to see, and the streak on the legs is fainter. The side of the thorax often has whitish markings.

Heptagenia sulphurea (TWO-TAILS)

DUN

Yellow May dun, male
(Plate IV. 47)

EYES: Dark blue becoming paler with age.
WINGS: Yellow with brown veins.
BODY: Dorsum—brownish yellow.
Venter—pale yellow.
LEGS: Yellow with greyish feet.
TAILS: Grey.
SIZE: Medium-large.

DESCRIPTIONS OF UPWINGED FLIES 69

Yellow May dun, female
(Plate IV. 46)

EYES: Black-green becoming paler with age.
WINGS: Pale yellow with yellow veins.
BODY: Very pale yellow varying to a deeper yellow.
LEGS: Pale yellow with greyish feet.
TAILS: Grey.
SIZE: Medium-large.

Remarks: Similar to the Yellow Evening dun but larger, and of a more vivid yellow colour, with two tails; the Yellow Evening dun has three tails. Both dun and spinner have upright hindwings with a costal projection. (Fig. 29)

SPINNER

Yellow May spinner, male
(Plate VI. 74)

EYES: Dark blue becoming pale blue with age.
WINGS: Transparent with dark brown veins, leading edge smoky grey.
BODY: Dorsum—dark olive-brown.
Venter—light golden olive-brown to yellow-green.
LEGS: Golden brown.
TAILS: Pale brown ringed dark brown.
SIZE: Medium Large.

Remarks: A beautiful golden brown bodied spinner with heavily veined wings. Blue eyes and exceptionally long forelegs. Similar in appearance to the Olive Upright male spinner.

Yellow May spinner, female
(Plate VI. 73)

EYES: Blue becoming paler with age.
WINGS: Transparent, but with a pale yellow area along leading edge. Veins dark brown.
BODY: Dorsum—pale olive-yellow.
Venter—pale yellow with last two segments yellow.
LEGS: Dark olive-yellow.
TAILS: Dark grey ringed brown.
SIZE: Medium-large.

Remarks: A large yellow to pale yellow spinner with heavily veined wings.

Heptagenia lateralis (TWO-TAILS)

DUN

Dusky Yellowstreak, male
- EYES: Black-brown.
- WINGS: Very dark grey.
- BODY: Dark greyish brown.
- LEGS: Dark brown-olive.
- TAILS: Greyish.
- SIZE: Medium-large.
- Remarks: The duns have an overall drab appearance, the predominant colour being grey.

Dusky Yellowstreak, female
- EYES: Black.
- WINGS: Very dark grey.
- BODY: Dark greyish brown with venter slightly lighter.
- LEGS: Dark brown-olive.
- TAILS: Greyish.
- SIZE: Medium-large.
- Remarks: Both dun and spinner have upright hindwings *with costal projection* (Fig. 11) and have a conspicuous yellow mark or streak each side of the thorax in front of forewing root.

SPINNER

Dusky Yellowstreak spinner male
- EYES: Dark brown.
- WINGS: Transparent with brownish veining along leading edge.
- BODY: Dark brown-olive with reddish ringing, paler underneath.
- LEGS: Brown-olive.
- TAILS: Brown.
- SIZE: Medium-large.

Dusky Yellowstreak spinner, female (Plate II. 27)
- EYES: Dark brown.
- WINGS: Transparent with brown veining along leading edge.
- BODY: Dorsum—brown-olive with reddish rings.
  Venter—drab-olive, last three segments orange-brown.
- LEGS: Brown-olive with red-brown patches.
- TAILS: Brown.
- SIZE: Medium-large.

31. **LARGE GREEN DUN—FEMALE**
    Ecdyonurus insignis

32. **AUTUMN DUN—MALE**
    Ecdyonurus dispar

33. **MARCH BROWN DUN—FEMALE**
    Rhithrogena haarupi

34. **MARCH BROWN DUN—MALE**
    Rhithrogena haarupi

35. **TURKEY BROWN DUN—FEMALE**
    Paraleptophlebia submarginata

36. **MARCH BROWN FEMALE SHOWING START OF TRANSPOSITION FROM DUN TO SPINNER**

37. **THE WELSMAN'S BUTTON**
    Sericostoma personatum

38. **SAND FLY (SEDGE FEMALE)**
    Rhyacophila dorsalis

39. **SILVER SEDGE (Vide HARRIS)**
    Lepidostoma hirtum

40. **BLACK SEDGE**
    Silo nigricornis

PLATE III.   DUNS AND SEDGE-FLIES × 1·6

Paraleptophlebia submarginata (THREE-TAILS)

DUN

Turkey Brown, male

EYES: Very dull red.
WINGS: Fawn with dark blackish veins, giving wings a mottled appearance.
BODY: Very dark blackish-brown.
LEGS: Dark brown.
TAILS: Dark grey-brown.
SIZE: Medium-large.

Turkey Brown, female (Plate III. 35)

EYES: Black tinged reddish.
WINGS: Fawn with dark blackish veining, giving wings a mottled appearance.
BODY: Very dark blackish brown, tinged olive and red.
LEGS: Very dark olive-brown.
TAILS: Dark grey-brown.
SIZE: Medium-large.
Remarks: A very dark brown-bodied dun, with mottled wings and upright hindwings without costal projection. (Fig. 12) There is a distinct clear patch in the centre of each forewing.

SPINNER

Turkey Brown, male spinner (Plate VIII. 101)

EYES: Very dark red.
WINGS: Transparent, veining along leading edges pale brown.
BODY: Dorsum—translucent brown with pale streaks ringed cream with last three segments dark brown.
Venter—similar to dorsum but paler.
LEGS: Very dark brown.
TAILS: Very pale brown ringed black.
SIZE: Medium-large.
Remarks: On the underside of each body segment there is a distinct oval brown mark (on male only).

Turkey Brown, female spinner (Plate VI. 78)

EYES: Grey-brown.
WINGS: Transparent, veining pale brown.
BODY: Dorsum—dark brown ringed pale brown.
Venter—similar to dorsum but paler.
LEGS: Forelegs dark brown, remainder brown.
TAILS: Pale brown ringed faintly black.
SIZE: Medium-large.

Paraleptophlebia cincta (THREE-TAILS)

DUN

Purple dun, male

EYES:	Dull reddish brown.
WINGS:	Blackish grey.
BODY:	Grey-black tinged purple.
LEGS:	Dark olive-brown tinged reddish.
TAILS:	Grey.
SIZE:	Medium.

Purple dun, female

EYES:	Dull green.
WINGS:	Blackish grey.
BODY:	Very dark brown tinged purple.
LEGS:	Pale olive-brown tinged reddish.
TAILS:	Grey.
SIZE:	Medium.

Remarks: At a cursory glance very similar to a large-sized Iron Blue but with an overall purple tinge.
  Upright hindwings without costal projection. (Fig. 12)

SPINNER

Purple spinner, male

EYES:	Brown-black.
WINGS:	Transparent with faint brown veining.
BODY:	Dorsum—translucent white, often tinged with mauve, last three segments purple-brown. Venter—similar but last three segments purple-grey.
LEGS:	Whitish tinged brown-red or mauve.
TAILS:	White.
SIZE:	Medium.

Purple spinner, female

EYES:	Black.
WINGS:	Transparent with faint brown veining.
BODY:	Dorsum—brownish with a purple tinge. Venter—similar but paler.
LEGS:	Very pale brown tinged with mauve.
TAILS:	Yellowish.
SIZE:	Medium.

Habrophlebia fusca (THREE-TAILS)

DUN

Ditch dun, male

- EYES: Dark reddish.
- WINGS: Dark grey with brownish veining.
- BODY: Dorsum—dark grey-olive.
- LEGS: Dark grey-olive.
- TAILS: Dark grey.
- SIZE: Medium.

Ditch dun, female

- EYES: Dark green.
- WINGS: Dark grey with brownish veining.
- BODY: Dorsum—dark brown-olive.
  Venter—very dark pink.
- LEGS: Grey-olive.
- TAILS: Dark grey.
- SIZE: Medium.

Remarks: The female is similar in appearance to the female B.W.O. but a little smaller and the body is a much darker colour.

The wings are very sparsely veined and the upright hindwings have a prominent costal projection half-way along the leading edge. (See Fig. 24)

SPINNER

Ditch spinner, male

- EYES: Dark reddish.
- WINGS: Transparent with light brown veining.
- BODY: Translucent reddish brown, last three segments brown.
- LEGS: Watery olive with brownish coloration around base of each femur joint.
- TAILS: Pale brownish grey.
- SIZE: Medium.

Ditch spinner, female

- EYES: Pale green.
- WINGS: Transparent with light brown veining.
- BODY: Dorsum—reddish brown, last segment yellowish.
  Venter—pale brown-red with edge of each body segment olive-green.
- LEGS: Pale watery olive, with brown around base of each femur joint.
- TAILS: Pale brownish grey.
- SIZE: Medium.

Leptophlebia vespertina (THREE-TAILS)

DUN

Claret dun, male

- EYES: Very dark red-brown.
- WINGS: Very dark grey with hindwing of a much paler shade.
- BODY: Dorsum—very dark black-brown, last three segments with a claret tinge.
  Venter—grey-black.
- LEGS: Dark brown-black.
- TAILS: Dark grey-brown.
- SIZE: Medium-large.

Claret dun, female (Plate II. 24)

- EYES: Black.
- WINGS: Very dark grey, hindwings pale buff colour.
- BODY: Dorsum—dark brown almost black with a claret tinge.
  Venter—brown with a claret tinge.
- LEGS: Dark brown.
- TAILS: Dark brown.
- SIZE: Medium-large.

Remarks: Both male and female duns closely resemble the Iron Blue, but are larger and have a more claret-brown body. The paler colour of the upright hindwings can be a helpful feature in identification; they have no costal projection. (See Fig. 12)

SPINNER

Claret spinner, male (Plate VIII. 97)

- EYES: Very dark red-brown.
- WINGS: Transparent lightly veined brownish.
- BODY: Dorsum—dark brown with a claret tinge.
  Venter—similar but paler.
- LEGS: Dark black-brown.
- TAILS: Pale brown ringed with red.
- SIZE: Medium-large.

Claret spinner, female (Plate VIII. 96)

- EYES: Black.
- WINGS: Transparent lightly veined pale brown.
- BODY: Dorsum—brown with a claret shade.
  Venter—similar but paler.
- LEGS: Brown to pale brown.
- TAILS: Pale brown lightly ringed with black.
- SIZE: Medium-large.

Leptophlebia marginata (THREE-TAILS)

DUN

Sepia dun, male
(Plate IV. 43)

EYES: Very dark red-brown.
WINGS: Pale fawn heavily veined brown.
BODY: Dorsum—dark sepia-brown.
Venter—grey-brown.
LEGS: Forelegs dark brown, remainder dark olive-brown.
TAILS: Brown.
SIZE: Medium-large.

Sepia dun, female
(Plate IV. 42)

EYES: Brown.
WINGS: Pale fawn, heavily veined brown.
BODY: Dorsum—dark sepia-brown.
Venter—pale brown.
LEGS: Forelegs brown, others olive-brown.
TAILS: Brown.
SIZE: Medium-large.

Remarks: The three tails of this fly are spread well apart. They have upright hindwings, without costal projection. (See Fig. 28)

SPINNER

Sepia spinner, male
(Plate VI. 80)

EYES: Dark red-brown tinged dull green.
WINGS: Transparent, veined light brown.
BODY: Dark brown ringed straw.
LEGS: Forelegs dark brown, others golden brown.
TAILS: Very long and dark brown.
SIZE: Medium-large.

Sepia spinner, female
(Plate VIII. 98)

EYES: Dark brown.
WINGS: Transparent, veined light brown.
BODY: Dorsum—dark red-brown.
Venter—pale red-brown.
LEGS: Forelegs dark brown, others golden brown.
TAILS: Very long and dark brown.
SIZE: Medium-large.

Remarks: The forewings of these spinners have a dark smoky black patch along the top leading edge.

Ecdyonurus dispar (TWO-TAILS)

DUN

Autumn dun, EYES: Deep greenish brown.
male WINGS: Grey, heavily veined black giving wings
(Plate III. 32) a slightly mottled appearance, with veins
along leading edge brown.
BODY: Dorsum—yellow-olive with dark brown
diagonal bands along sides.
Venter—brown-olive.
LEGS: Dark brown-olive with feet darker.
TAILS: Very dark grey.
SIZE: Large.

Autumn dun, EYES: Brown-black.
female WINGS: Pale fawn with heavy blackish veins, giving
wings a slightly mottled appearance.
BODY: Dorsum—yellow-olive or pale olive-brown.
Sides touches of red-brown.
Venter—pale olive-brown.
LEGS: Dark brown-olive.
TAILS: Very dark grey.
SIZE: Large.
Remarks: A rather large fly with heavily veined wings and a beige-coloured body. The forelegs of the male are very long.
    Both dun and spinner have large upright hind-wings, with a costal projection. (See Fig. 11)

SPINNER

Autumn spinner, EYES: Dark brown-black.
male WINGS: Transparent but with veining very distinct
(Plate VII. 95) and brownish.
BODY: Dorsum—dark brown-red, almost maho-
gany colour with blackish joinings.
Venter—dark olive with last two segments
yellowish.
LEGS: Brownish.
TAILS: Very dark grey-brown.
SIZE: Large.

41. MAYFLY DUN-FEMALE
Ephemera danica
x¾

42. SEPIA DUN-FEMALE
Leptophlebia marginata
x1

43. SEPIA DUN-MALE
Leptophlebia marginata
x1

44. LARGE BROOK DUN-FEMALE
Ecdyonurus torrentis
x1

45. LARGE BROOK DUN-MALE
Ecdyonurus torrentis
x1

46. YELLOW MAY DUN-FEMALE
Heptagenia sulphurea
x1

47. YELLOW MAY DUN-MALE
Heptagenia sulphurea
x1

48. GRANNOM FEMALE
Brachycentrus subnubilus
x1.4

49. GROUSE WING SEDGE
Mystacides longicornis
x1.4

50. BLACK SILVERHORN
Mystacides azurea
x1.4

51. MARBLED SEDGE
Hydropsyche contubernalis
x1.4

52. BROWN SILVERHORN
Athripsodes cinereus
x1.4

53, 54 & 55. LARGE DARK OLIVE FEMALE IN THREE DIFFERENT STAGES OF MOULT FROM DUN TO SPINNER x1

PLATE IV. DUNS AND SEDGE-FLIES x VARIOUS

DESCRIPTIONS OF UPWINGED FLIES

Autumn spinner, female
(Plate VI. 77)

EYES: Dark greenish brown.
WINGS: Transparent but with veins a distinct brown-black colour.
BODY: Dorsum—reddish brown.
Venter—darker red-brown.
LEGS: Olive-brown.
TAILS: Very dark brown.
SIZE: Large.
Remarks: A large dark red bodied spinner, sometimes referred to as the Great Red spinner in common with the Late March Brown and the Large Brook spinner.

Ecdyonurus torrentis (TWO-TAILS)

DUN

Large Brook dun, male
(Plate IV. 45)

EYES: Dark brown tinged green.
WINGS: Pale fawn veined black, giving a mottled appearance. Leading edge yellowish.
BODY: Dorsum—pale olive-brown, banded diagonally red-brown along sides.
Venter—pale purple-brown.
LEGS: Forelegs dark brown, remainder dark olive.
TAILS: Purple-brown.
SIZE: Large.

Large Brook dun, female
(Plate IV. 44)

EYES: Brown-olive.
WINGS: Pale fawn, veined dark brown, giving a mottled appearance. Leading edge yellowish.
BODY: Dorsum—drab olive-brown, diagonally banded red-brown along sides.
Venter—purple-brown.
LEGS: Dark olive-brown, tinged purple.
TAILS: Purple-brown.
SIZE: Large to very large.
Remarks: Both male and female dun have mottled wings similar to the March Brown, with clear areas in the centre, but in addition they also have dark shaded bands across the forewings. They have upright hindwings, with a costal projection. (See Fig. 11)

SPINNER

Large Brook spinner, male
(Plate VI. 72)

EYES: Black with green tinge.
WINGS: Transparent with brown veining, leading edge yellowish.
BODY: Dorsum—dull olive-brown with diagonal purple-brown banding along sides.
Venter—pale purple-brown.
LEGS: Forelegs dark brown, others pale brown.
TAILS: Purple-brown.
SIZE: Large.
Remarks: Tails very long on male spinner.

Large Brook spinner, female
(Plate VI. 71)

EYES: Brown.
WINGS: Transparent veined black, leading edge yellowish.
BODY: Dorsum—dark olive-brown, diagonally banded red along sides.
Venter—purple-brown.
LEGS: Dark olive-brown, feet blackish.
TAILS: Purple-brown.
SIZE: Large to very large.
Remarks: In both dun and spinner the female is considerably larger than the male.
This female spinner has a distinctly reddish purple coloration overall. It is sometimes referred to in common with the Autumn Dun spinner and the Late March Brown spinner as the Great Red spinner.

Ecdyonurus insignis (TWO-TAILS)

DUN

Large Green dun, male

EYES: Dark greenish brown.
WINGS: Pale fawn with very heavy brown veining, giving a mottled appearance.
BODY: The whole of the body is a dark olive-green colour with diagonal brown bands along the sides.
LEGS: Dark olive. Feet brown.
TAILS: Dark grey-black.
SIZE: Large.

Large Green dun, female
(Plate III. 31)

EYES: Dark greenish brown.
WINGS: Pale fawn with heavy brown-black veins, giving a mottled appearance.
BODY: The body of the female is very similar to the male, although the coloration is a little lighter.
LEGS: Dark olive, feet darker.
TAILS: Very dark grey.
SIZE: Large.
Remarks: Similar in appearance to the Autumn Dun except that the body is a greenish shade. The wings have a distinct clear patch in the centre which is almost clear of veins. This feature is apparent in both male and female. They have upright hindwings with a costal projection (see Fig. 11) and the under abdominal sections of both dun and spinner are marked in a distinctive way. (See Fig. 15)

SPINNER

Large Green spinner, male

EYES: Olive-green with brown flecks.
WINGS: Transparent with very heavy dark brown veins.
BODY: Olive-green with dark brown diagonal bands along the sides.
LEGS: Dark grey-olive.
TAILS: Dark grey-black.
SIZE: Large.
Remarks: The spinners have a distinctive smoky black shaded area along the top leading edge of the forewings. The males have exceptionally long tails.

Large Green spinner, female

EYES: Olive-green with brown flecks.
WINGS: Transparent with heavy dark brown veins.
BODY: Olive-green with dark brown diagonal bands along the sides.
LEGS: Dark olive.
TAILS: Dark grey-brown.
SIZE: Large.
Remarks: This spinner sometimes has a brown triangular-shaped mark on the middle top of each body segment, but it is not constant.

E. Notata   *E. Insignis*

Fig. 15. Showing constant black markings on ventral segments of the abdomens of *Ephemerella notata* and *Ecdyonurus insignis*.

Ecdyonurus venosus (TWO-TAILS)

The Late March Brown  The male and female dun closely resemble the male and female March Brown in size and colour respectively, but they lack the pale areas in the centre of the forewings.

The male spinner is very similar to the Autumn spinner (*E. dispar*), but is slightly larger in size and a little paler in colour. The female Great Red spinner is also similar to the female of the Autumn spinner, but is larger and redder in the body.

WING VENATION OF THE UPWINGED FLIES

The following sketches and photographs of the wings of the Ephemeroptera are included in order to assist the angler/entomologist to identify specimens more positively; this can be done by studying the venation, or veining, of the wings, particularly in the case of some of the more difficult types, where it is an essential point of reference.

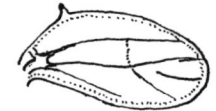

Fig. 16. *B. pumilus*.

Fig. 17. *B. niger*.

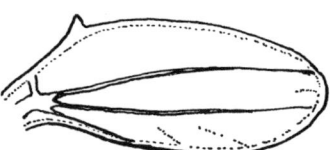

Fig. 18. *B. rhodani*.

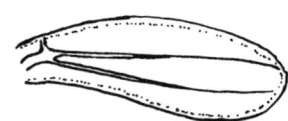

Fig. 19. *B. atrebatinus*.

Fig. 20. *B. vernus.*

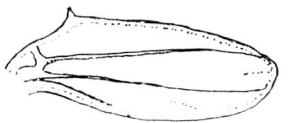

Fig. 21. *B. scambus.*

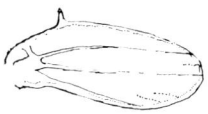

Fig. 22. *B. bioculatus.*

Fig. 23. *B. buceratus.*

Fig. 24. *H. Fusca.*

Fig. 25. *C. luteolum.*

Fig. 26. *C. pennulatum.*

Fig. 27. *E. ignita.*

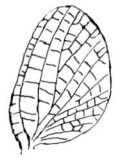

Fig. 28. *L. marginata.*

Fig. 29. *H. sulphurea.*

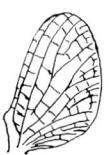

Fig. 30. *E. venosus.*

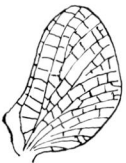

Fig. 31. *R. semicolorata.*

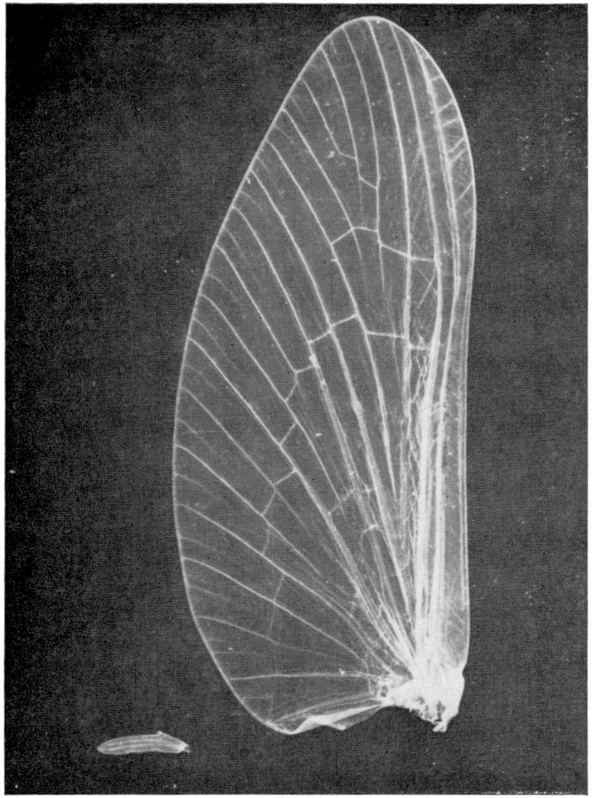

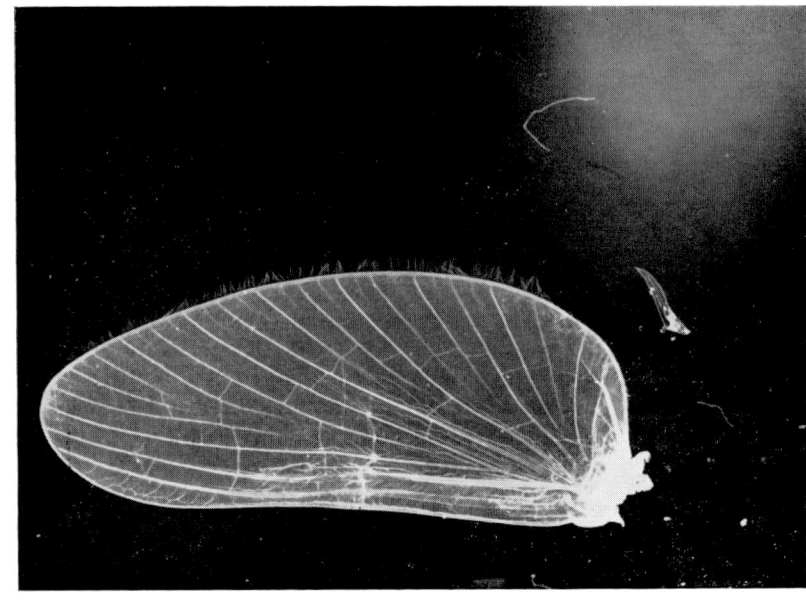

Fig. 33. *C. Luteolum* ×10. Forewing and small narrow hindwing of subimago female. Note single intercalary veins.

Fig. 32. *C. pennulatum* ×10. Forewing and small narrow hindwing of subimago female. Note single intercalary veins.

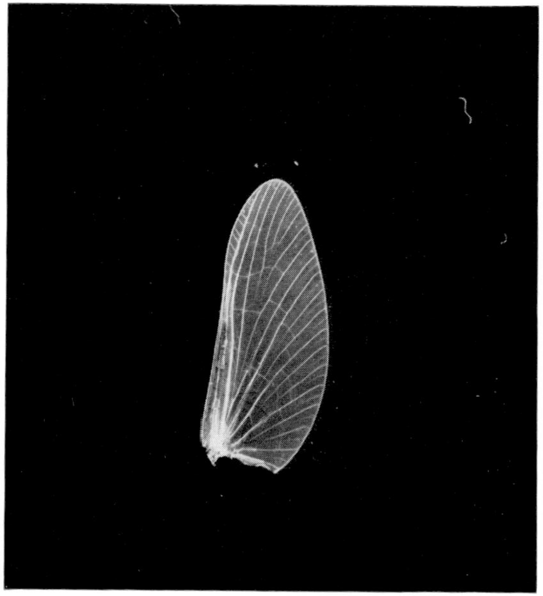

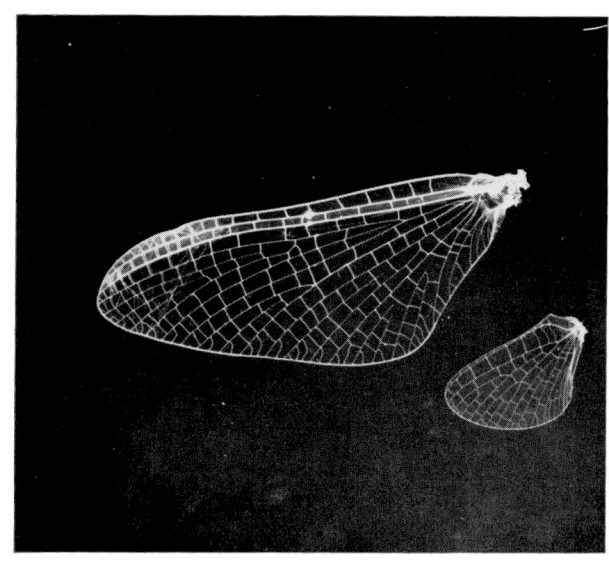

Fig. 34. *P. pseudorufulum* × 5. Forewing of subimago female. Note six to eight cross veins in Pterostigmatic area.

Fig. 35. *H. sulphurea* × 5.5. Forewing and upright hindwing of imago male.

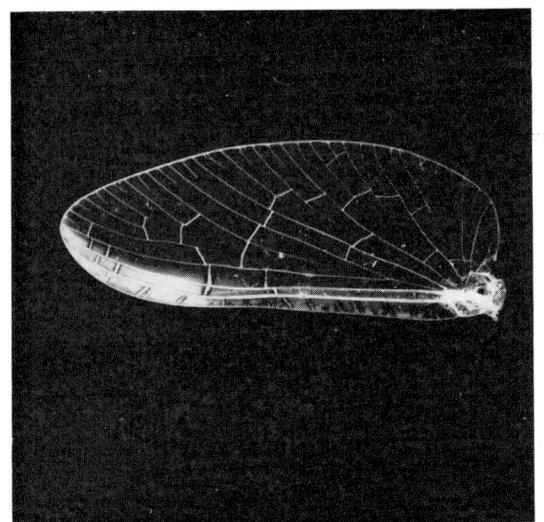

FIG. 37. *C. dipterum* ×5.5. Forewing of imago female. Note three to five cross veins in Pterostigmatic area.

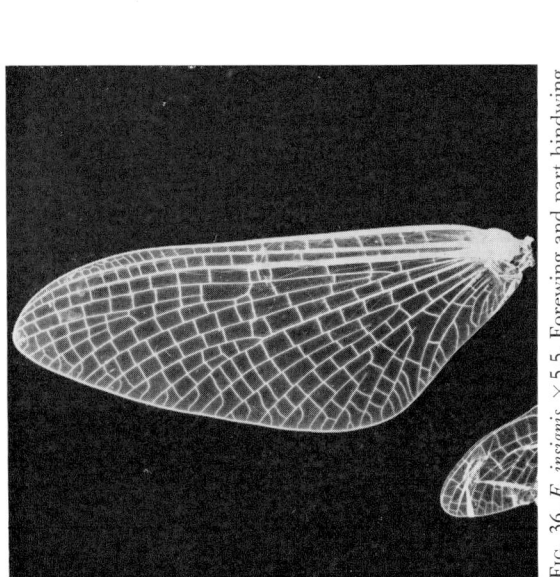

FIG. 36. *E. insignis* ×5.5. Forewing and part hindwing of subimago female.

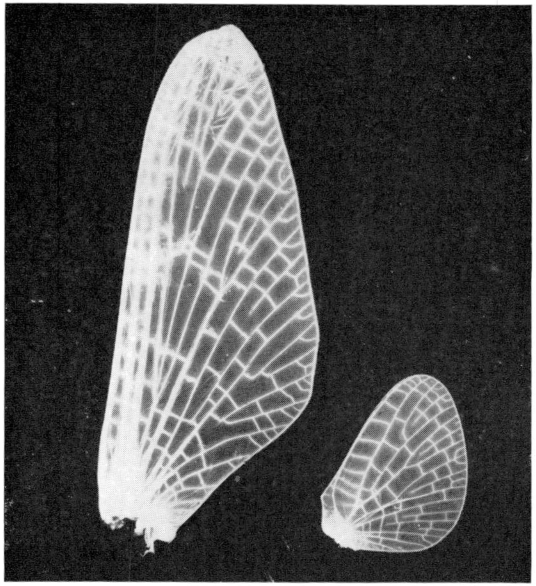

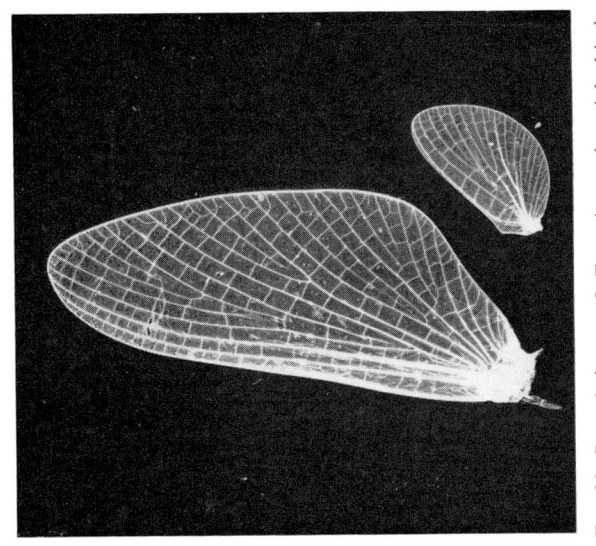

FIG. 39. *R. semicolorata* ×5. Forewing and upright hindwing of subimago female.

FIG. 38. *R. haarupi* ×5. Forewing and upright hindwing of subimago male.

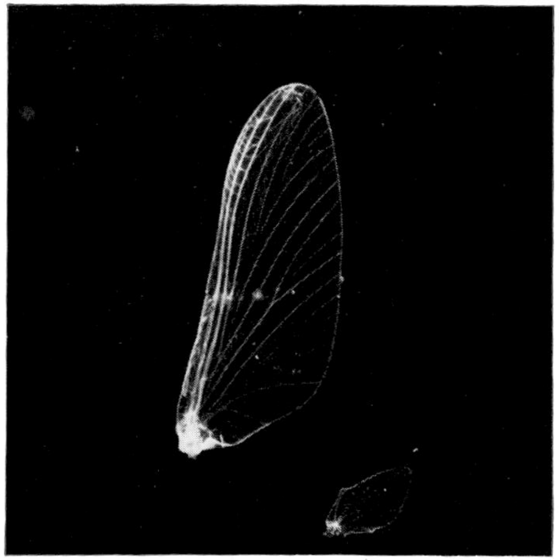

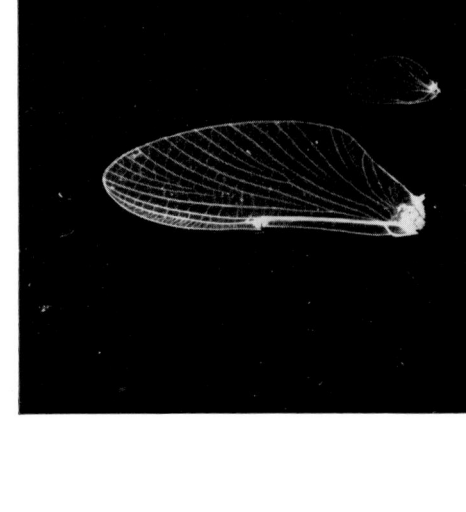

Fig. 40. *E. ignita* × 5. Forewing and upright hindwing of imago female.

Fig. 41. *L. vespertina* × 5. Forewing and upright hindwing of imago female.

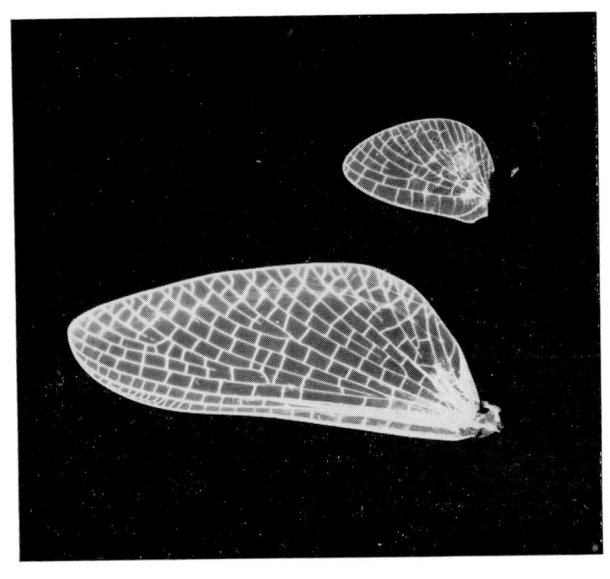

Fig. 43. *E. dispar* × 5. Forewing and upright hindwing of subimago male.

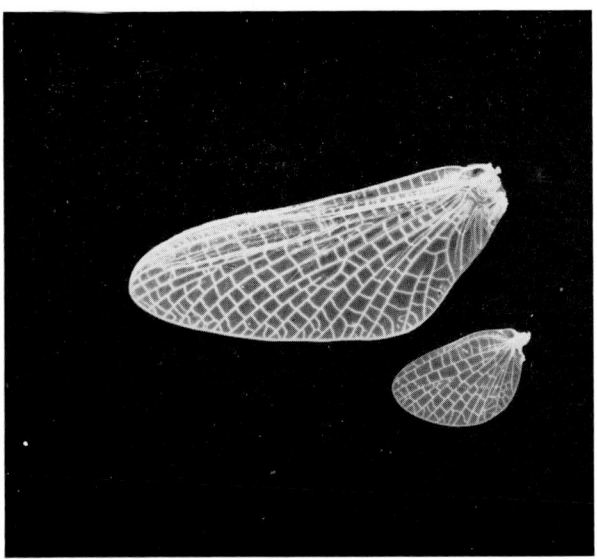

Fig. 42. *L. marginata* × 5. Forewing and upright hindwing of subimago male.

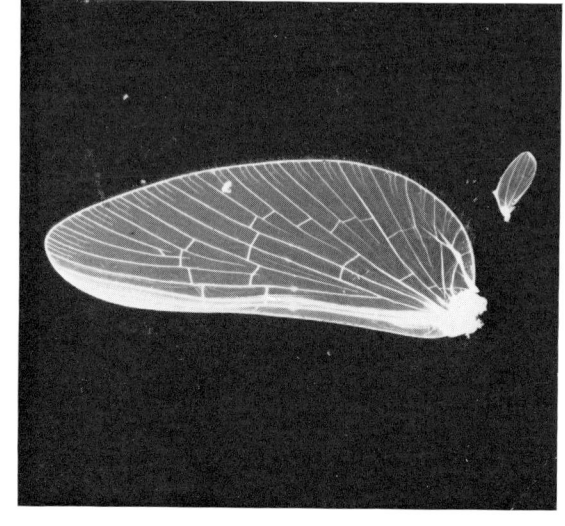

Fig. 45. *B. rhodani* × 5.5. Forewing and small oval hindwing of subimago female. Note double intercalary veins.

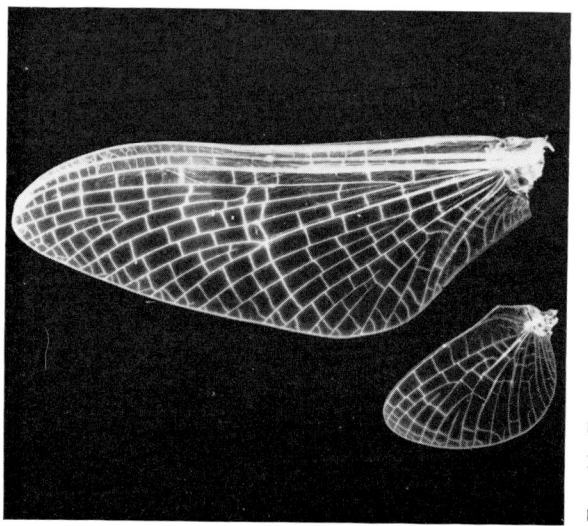

Fig. 44. *E. torrentis* × 5. Forewing and upright hindwing of subimago male.

F

CHAPTER V

FISHING INFORMATION ON THE UPWINGED FLIES AND THEIR DISTRIBUTION

In the preceding chapters the details given have been strictly factual, but in this chapter every endeavour has been made to include hints and views that could be helpful to the angler.

All aquatic flies live the greater part of their lives under water, and in the course of evolution have adapted themselves to the environment most suited to the particular species. It will be appreciated therefore that the under-water forms of these flies (nymphs) vary considerably, both in actual physical shape and characteristic features evolved to suit their aquatic mode of life. Some nymphs have adapted themselves to rough stony streams, such as those of the North, and live the best part of their lives clinging with their flat bodies to stones. Others prefer slower water and live in little tunnels which they excavate in the bed of the stream. Various families of nymphs found in our chalk or limestone rivers live largely in the weed beds which abound in most of these waters. Some of them have a preference for alkaline water, others for slightly acid water. Some thrive in the highly oxygenated water of mountain streams, others in sluggish lowland rivers.

The study of insects in relation to their surroundings is known as the science of ecology, and from it we often can gather clues regarding the distribution of insects which can assist us to identify them when an element of doubt exists. To clarify this point further, let us imagine we are having a pleasant week's fishing holiday on an unfamiliar water. We may possibly rely on local information about the flies that are likely to be encountered, but this can sometimes prove misleading, and in any event if we have an inquiring mind we will probably want to identify the flies ourselves. Shortly after arrival at the water a few duns start to hatch out, and several trout and grayling move to them. We quickly catch one of the duns and, searching our memory (or perhaps even this volume), we come to the conclusion that our specimen is either a March Brown or a Turkey Brown. The latter has three tails and the former only two, so identification should normally be no problem, but in this case we will assume that the Turkey Brown has lost one of its tails, which occasionally happens. If so, we would probably conclude wrongly that our specimen is a March Brown, but a knowledge of their habits and distribution tells us that this is highly improbable: March Browns inhabit swift-flowing rivers with stony bottoms, and the river we are fishing is rather sluggish with a muddy bottom, much more suited to the Turkey Brown. Although this is an extreme case, it should help to illustrate the point being made, which relates to the preference of certain species for different habitats. Our study of ecology might even help us to decide where one may find different species of flies in a particular river, the more so when one happens to be fishing a stretch with varying water characteristics that support different insect species. A typical example is as follows.

The Blue Winged Olive, one of the most common chalk stream flies, often emerges at dusk, and an angler will gain an advantage if he can recognize in advance the most likely part of the river where this is likely to occur. Occasionally, especially on a bleak day, spinners or indeed any other insects are in short supply and the best chance of some activity may be found in a rise to a hatch of the B.W.O. An angler knowing a little about the habitat of various species will be aware of the fact that such an event is most likely to take place below fast water where it runs over shallows, and will station himself accordingly.

The Mayfly (*Ephemera danica* or *vulgata*) (Plate IV. 41)

This is the largest of the Upwinged flies, and is often referred to as the Greendrake. Probably more has been written on this particular fly than all the other upwinged species put together, so therefore it is not felt necessary to expound at too great a length. Even the merest novice fly-fisherman must be familiar with this distinctive-looking insect. With its heavily veined wings, cream-coloured body, very large size and three tails, it can hardly be confused with any of the other Upwinged flies. The difference between the two species, *danica* and *vulgata*, is so slight that it is of little consequence to the angler. However, for the angler entomologist, *E. danica* is probably the more common of the two and occurs in the faster flowing rivers and lakes. *E. vulgata* seems to favour the more sluggish rivers with muddy bottoms. For the record, there is also a third species *E. lineata*, but it is very rare in these islands. For positive identification of these three species the only certain way is to refer to the sketches (Fig. 46) of the markings of the underside of the body segments. Mayflies are abundant and common in most parts of the British Isles including Ireland, but they only occur in local areas in the North of England and Scotland. They have a comparatively short season, usually lasting about two weeks, in either late May or early June. Hatches tend to build up to a crescendo during this period, culminating usually in two or three days of really prolific hatches at the end of the fourteen days, after which they rapidly thin out. During these few days the emergence normally commences about midday, often continuing till early evening. However, on some occasions I have experienced a delay in the main hatch until four or five o'clock. Unfortunately the colossal hatches of this fly that used to be prevalent seem to have diminished, perhaps due to the ever-increasing effects of pollution in many of our rivers, or a natural cycle tendency.

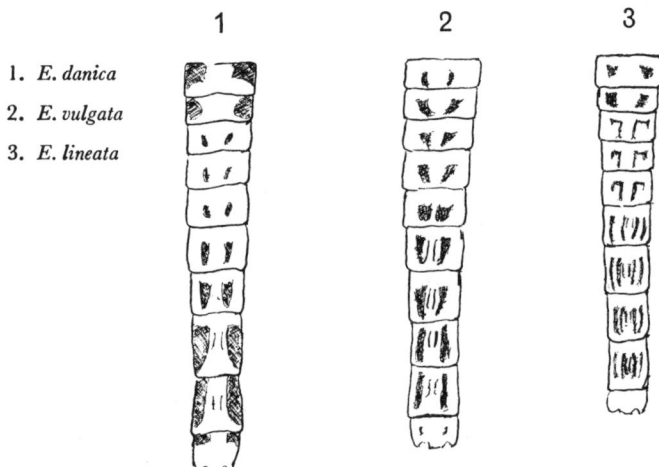

1. *E. danica*
2. *E. vulgata*
3. *E. lineata*

Fig. 46. Showing difference between constant black markings on dorsum segments of abdomens of Mayflies.

The Spent Gnat

This is the popular name given to the female spinner of the Mayfly as she drifts along, dying, on the surface after laying her eggs. The body of this spinner is a pale-cream colour, and the wings although transparent have a distinct bluish tint. While ovipositing the female spinner will often alight on the water and drift downstream for short periods, and at this stage she is referred to as a Grey Drake, the artificial pattern of which is particularly good at this period. The male spinners of the Mayfly are also worthy of mention as they often fall on to the water after mating and are carried along to the waiting fish. In appearance they have dark legs and thorax, with a creamy white body, and the three tail segments of the body are brown. The transparent wings are tinged quite distinctly with dark patches. A fall of Mayfly spinners is always eagerly awaited by fly-fishermen, as the trout and grayling seem to prefer the spinner to the hatching dun, and take them readily. Also, of course, it is much easier to deceive the fish with an artificial pattern representing the spinner. A useful indication of whether or not there is likely to be a fall of spinners in the evening can often be determined by looking for swarms of the male spinners. These normally form in the vicinity of trees or bushes along the river banks in the late afternoon or early evening, and if the swarms are absent or sparse by mid-evening it is doubtful whether any appreciable fall will occur.

The Large Dark Olive (*Baëtis rhodani*) (Plate II. 16 and 17)

Without a doubt this is one of our most common and widespread Upwinged flies and is found on many of our rivers. Although odd hatches sometimes occur during the summer, and I once observed such an event in August on a small Scottish stream, yet for the most part they appear in quantity only in the late autumn or early spring. Indeed, when the trout season opens they are often the only duns on the water, when they can be observed on cold still days floating along the surface for a hundred yards or more before their wings are sufficiently dry to support flight. Fortunately for their survival the fish at that time of the year lie deep, and are not always interested in surface food; the absence of many habitual birds of the streams also aids their preservation. This largish conspicuous Ephemerid makes its home on all types of waters, acid and alkaline, fast large brawling rivers like the Usk, slow rivers like the Upper Thames, and small streams and brooks. The colour, though typically dark olive, varies somewhat between male and female, the former being a little darker than the latter. I am given to understand that in some parts of the country the body colour is darker than in others, and I am indebted to Mr. Thomas Clegg, the well-known fly dresser, for the information that in some parts of Scotland the body of the dun is a dark lead colour, and that the wing veins are almost black.

To the trout fly-fisherman it is an important insect mostly in the month of April, as the autumn hatches occur just at about the time when the season is at its close. Grayling fishers however may find it a very useful fly to imitate.

Of the many good artificial patterns which represent this dun, I must mention the Imperial, which was invented in 1961–62 by Major Oliver Kite, and for which I have given the full dressing later in this book. Its success may be due to the purple tying silk which tinges the heron herl of the body in a most realistic way.

The Large Dark Olive Spinner (Plate VII. 81)

The Large Red spinner, as the female is so aptly called, is seldom seen on the water in any quantity by the average fly-fisher. The reason for this is rather hard to define, as this spinner must certainly return to the water to lay her eggs. Perhaps mating and depositing of the eggs are spread over most of the day instead of during a confined period in the evening, as in the case of many other species of the Upwinged flies of summer. Of course it is possible they may not commence laying their eggs until after dusk, or deposition may be concealed from us by some other method.

The Iron Blue (*Baëtis pumilus* or *niger*) (Plate I. 1 and 2)

Of the two species of this fly, the former is more widely distributed than the latter, but from the angler's viewpoint they are so alike that there is no need to differentiate. They are fairly widespread over the whole of our islands, except in the South-eastern part of England. They are found in much the same habitat as the previous species, and are also common to both acid and alkaline waters. *B. niger* seems to have a preference for rivers with a substantial weed growth, and *B. pumilus* is more common on small stony streams. Hatches are most prolific in May and early June and again during September and October, but they occasionally occur in smaller quantities at odd intervals throughout the summer. This species, unlike most of the other Baëtids, seldom hatches out regularly and in some localities is seen only infrequently. Another interesting facet of its nature is that hatches very often occur in the most adverse weather conditions. The wetter or windier the day the more likelihood of a good hatch. On several occasions during the past few years I have had successful days in early May on the River Itchen with an imitation of this fly, and always conditions have been similar. My arrival at the waterside has been greeted by overcast skies accompanied by blustery wind and often stinging rain. On such days, hatches of the more common Olives are either sparse or entirely absent, and expectations of sport are usually low. Sometimes, however, I fortunately observe a slow trickle of these tiny dark flies drifting down on the surface like little blobs of ink, often starting off rather sparsely in the morning and

56. THE CAPERER
Halesus radiatus

57. LARGE CINNAMON SEDGE
Potamophylax latipennis

58. THE GREY OR SILVER SEDGE
Odontocerum albicorne

59. THE GREY FLAG
Hydropsyche pellucidula

60. CINNAMON SEDGE
Limnephilius lunatus

61. BROWN SEDGE
Anabola nervosa

62. THE ALDER
Sialis Spp.

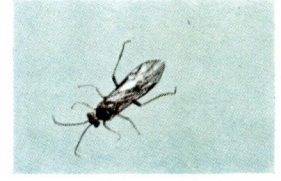

63. THE SAW FLY

64. THE WATER LOUSE
Asellus Spp.

65. SHRIMP
Gammarus pulex

66. LARVA OF NON-CASEMAKING CADDIS
Hydropsyche Spp.

67. LARVA OF CADDIS IN CASE

68. HAWTHORN FLY
Bibio marci

69. AUTUMN DUN-SPINNER FEMALE
PHOTOGRAPHED FROM UNDER WATER

70. NYMPHAL CASE SPLITTING AS DUN (*E. torrentis*) EMERGES AT THE SURFACE
Photography by courtesy of Dr. Michael Wade

PLATE V.   SEDGE-FLIES AND OTHER FAUNA x 1

gradually increasing in volume towards the middle of the day. On several such outings, I have witnessed these little flies hatching in incredibly large numbers; in the eddies, at the tails of pools or edges of hatches, the surface of the water has appeared almost dark blue with hundreds of dead and dying duns that have failed to become airborne. Although the dun when appearing in quantities is fairly easy to identify, it can often be missed due to its very small size, when sparse hatches coincide with those of other larger duns. Other points of interest are that flies of the spring generation seem to be a little larger and a little darker than those of the autumn generation.

The Iron Blue Spinner (Plate VII. 83 and 86)

The popular name for the female spinner of this species is the Little Claret spinner. The male spinner, known as the Jenny spinner, is one of the few male spinners sometimes found on the water in sufficient quantities to arouse the interest of the fish, but it seems to be generally accepted that trout are not particularly keen on it. It is a very handsome little fly, the transparent gauzy wings contrasting sharply with the jet-black thorax. Its body is pearly-white, and the last three segments are a dark orange-brown. To complete the picture, its two long tails are almost white. The female or Little Claret spinner is drab in comparison, with a rather dark claret-brown body, but to the fish it is a more important fly, for when it is on the water trout will feed on it avidly; unfortunately falls of this particular spinner vary in frequency from river to river, due of course to the different densities in population of the species. However, if one is on the river on a warm day following several cold days, when hatches of these duns are likely but swarming and mating unlikely, one may very fortunately witness a fall of spinners of considerable proportions. The best months of the year to encounter these female spinners as they lie spent in the water after egg-laying, are late May, June, September and early October, A further interesting feature to note is that the spinners of this species can be observed in the water at any time during the day, as they lay their eggs in the mornings and afternoons as well as in the evenings. They are difficult to see because of their small size, and the fact that they are usually carried along under the surface film.

The Medium Olive (*Baëtis vernus*, *B. tenax* and *B. buceratus*)
(Plate I. 3 and 6)

The two latter species can be regarded as identical with *B. vernus*, as the differences are so slight; in fact most experts believe that *B.*

tenax is the same insect as *B. vernus*, which is widespread over the whole of the country, while *B. buceratus* is confined to the southern part of England and Wales. They first appear in mid-May and hatches, often prolonged, occur usually during the late morning or early afternoon on most days throughout the early summer and again in September. The duns are an olive colour, but appear to be a creamy colour in flight as they ascend from the surface of the water after hatching, due no doubt to the pale yellow-olive colour of the underside of the body. They are an important angling fly on many of our rivers, especially the chalk streams. They are not particularly fussy as to habitat and are to be found in most stretches of rivers, slow or fast.

The Medium Olive Spinner (Plate VII. 84)

The female is commonly known as the Red spinner, although it is certainly not always red in colour, as I have recorded specimens from yellow-brown to a dark red-brown. It should be explained that most spinners tend to darken with age. This particular species is often yellow-brown when freshly transposed, but quickly darkens, and after mating when fully spent is sometimes of a dead leaf colour. These spinners are usually to be found on the water in considerable numbers in areas where they occur, and falls often commence quite early in the evening. The male spinners are very similar to the males of the Large Dark Olive spinner, although a little smaller, and in some cases the body colour is a dark red-brown as opposed to pale olive-brown. These male Medium Olive spinners often form large swarms along the banks of the rivers or a little way inland.

The Small Dark Olive (*Baëtis scambus*) (Plate I. 7 and 8)

This dun is fairly common and is found in most rivers in England, Wales and the West Country, but is scarcer in Scotland and Ireland. It is abundant on chalk or limestone streams and rivers, although it seems to prefer mountain streams. The Small Dark Olive, or July dun as it has been called in the past, is mainly to be seen during July and August, when it is usually on the water in large numbers. Hatches can be expected to start before noon and often a steady trickle will continue right through to about tea-time, but on several occasions I have observed good hatches of a late evening, particularly during the latter part of June. It is a tiny fly with a grey-olive body and smoky grey wings, and is a great favourite with both fish and fishermen, particularly on the chalk streams. The species contains some of the smallest of our Upwinged duns, apart from

Caënids, and in some cases they are even smaller than large specimens of these.

The Small Dark Olive Spinner (Plate VIII. 102 and 103)

The common name for the female spinner is the Small Red Spinner and although smaller, it is similar to the previous species. It varies considerably in colour from brown to the more common dark red-brown. These spinners often found spent on the water in the early evening, are usually difficult to see, as in many cases they float along just under the surface film. The reason for this is as follows: the insects of many species of the *Baëtis* genus often crawl under water to deposit their eggs. After this operation is completed, they endeavour to surface but are sometimes too weak to break through the surface film and are trapped under it. Some however do break through and get carried along on the top of the film, but those that are trapped underneath it are extraordinarily difficult to see from the bank. On many occasions during an evening rise I have been deceived by a fish feeding on these *Baëtis* spinners. I will present a typical experience of this nature in order to illustrate the importance of being on the alert at all times. On this particular occasion I was fishing that delightful little river the Wylye, and as dusk approached there was a good hatch of duns. After catching a few I found they were Blue Winged Olives and Pale Evening duns. Within minutes I spotted a fish rising, and first tried him with a B.W.O. pattern, but after several refusals I decided he must be taking the Pale Evening duns, so quickly changing to the appropriate pattern (Pale Evening Dun—Kite) I cast to him again, but was once more refused. I then proceeded to try with various B.W.O. patterns, but all to no avail. Perplexed, I retreated from my casting position to study the situation, and after a few minutes it became apparent that the rise form was in keeping with a fish feeding on spinners and not duns. I should of course have noticed this earlier had I not been obsessed with the idea that it was feeding on duns, although the difference in the rise is not always easy to identify, even to an experienced angler. Quickly dropping back to a spot where the current swept right in under the bank I peered directly down into the water, and there sure enough were Small Red spinners in plenty drifting along under the surface film. Returning, I mounted a small Pheasant-tail Spinner and hooked the fish on the second cast. Sad to relate, however, in the excitement I broke on the strike. The moral to be observed, after being refused by a fish apparently feeding on duns, is always to drop back down river to a spot where you can look directly down into the main current from the bank or a bridge for spinners. It does not always follow that spinners will be present, but when they are it may save a lot of time and wasted effort during the often all-too-short evening rise.

The male spinner is somewhat similar to the male Iron Blue (Jenny) spinner, except that the body is not such a translucent white and the thorax not so black. They quite often form swarms well back from the river banks in the shelter of surrounding bushes and trees.

The Pale Watery Dun (*Baëtis bioculatus*) (Plate I. 4 and 5.)

None of the Baëtids are easy to identify, and this particular species is probably the most difficult of all. This applies of course to the female, the male being very easy to identify with its yellow eyes. As the name implies, they are a pale watery olive colour. Unfortunately the colour is not very helpful for identifying the female, as both the Medium Olive (*vernus*) and the Small Dark Olive (*scambus*) are sometimes a similar colour. It is mainly confined to England and southern Wales, where it is very common, particularly on the chalk streams of southern England. Hatches occur in short bursts throughout most of the day from late May to October. In September and October fairly prolific hatches often occur during the early evening. As a matter of interest, since the beginning of the century, the name Pale Watery was applied to four different species: *B. scambus*, *C. luteolum* and *C. pennulatum*, and the species referred to here, *B. bioculatus*. The popular angling names for these four species now are: the Small Dark Olive, the Small Spurwing, the Large Spurwing and the Pale Watery.

The Pale Watery Spinner (also known as the Golden Spinner) (Plate VII. 87)

This seems to be a fairly descriptive name for the female spinner of the Pale Watery, which varies in colour from a pale golden brown to a golden olive. Despite the fact that the dun is relatively common, the spinner is not often seen spent on the water in any quantity. Falls usually seem to start fairly early in the evening, and in consequence are often spread over a couple of hours or more, and this could of course account for the apparent scarcity of them on some occasions. I have found Lunn's Yellow Boy a good artificial to use when this spinner is on the water, tied, of course, on a small hook, size 0 or 00.

The Yellow Evening Dun (*Ephemerella notata*)

This species is very similar in appearance and colour to the Yellow May dun; however, it can be readily identified as it has three tails and is also smaller. Hatches usually occur in the late evening, and the adult winged fly is most commonly seen during May and June. It is a rather uncommon species, and is somewhat local in its distribution. According to recent reports, it is more scarce now than it was a few years ago. The spinner is similar in appearance to the dun, but the wings are transparent with yellowish veining along the leading edges. Both dun and spinner have distinctive markings on the underside of the body segments (see Fig. 15).

The Blue Winged Olive (*Ephemerella ignita*) (Plate II. 22 and 23)

Without a doubt this is the most common, widespread and well known of all our Upwinged duns. With its bluish wings, three tails, olive-coloured body, it is probably the easiest of all our flies to recognize. The male is considerably darker than the female and often has a very brownish olive body. In the late autumn the body colour of both is often more of a rusty brown colour than olive. This species is found all over the British Isles in nearly all types of flowing water, and in many big lakes. It first appears in mid-June and continues right through to the end of the fishing season. In the early part of its season, hatches on some rivers are confined to the evenings only, often just as dusk approaches. Later on in August, when the days begin to shorten, hatches will often begin in the late afternoon or early evening, and on several occasions in recent years I have observed quite good hatches of this fly shortly after midday. It is a large and robust fly, and with experience the angler can soon learn to identify it from a fair distance, as it hatches and floats along on the surface. The reason for this is that the large forewings of this fly slope back slightly over the body, much more so than the wings of any of the Olives. Under favourable conditions hatches of this fly will occur over most parts of the river, but under unfavourable circumstances hatches will often be localized. A useful point to remember is that these usually take place below the well-aerated gravel or stony shallows, in fact below any areas of broken water on a river.

The Sherry Spinner (Plate VII. 93 and 94)

Although the colour range of these spinners, even of the individuals within the same swarm, varies from olive-brown to sherry-red to a distinct lobster-red, the most common shade is the sherry colour, whence the popular name. The females are often to be seen in vast swarms in the late evening, preparatory to egg-laying. As a matter of interest, it should be mentioned here that several leading authorities have always maintained that the female spinner carried her egg ball cupped under her tails, which she curls under for this purpose.

David Jacques, the well-known angler/entomologist, recently carried out a study on this subject, and his findings confirmed the opinion made several years earlier by the late Martin E. Mosely, whose views were not entirely accepted. The function of the tail in the extrusion process is as follows. When the egg ball is extruded from the vent of the female, the tails press the egg ball into the abdomen at a position between the ultimate and penultimate segments, after which they return to their original position—i.e. in a straight line to the rear—and the insect flies on her way with the egg ball held in position by the sticky nature of the egg mass. Retention is probably assisted by a pair of lobes which appear to be grasping, somewhat inadequately, the sides of the ball. The insect uses considerable force when she presses the egg ball into her abdomen, making a permanent dent in the body, which remains a prominent feature of her anatomy for the rest of her life, even after the eggs are released. A photograph of a Sherry spinner with the egg ball in position and tails normally outstretched appears on Plate II. 30 which proves the point beyond reasonable doubt. It is worth mentioning that male spinners of the B.W.O. are almost unique in that when they fall on the water in sufficient quantities, as they sometimes do, they arouse as much interest in the fish as do the females. On those occasions when I have observed this happen they have fallen prior to the spent females, and I must confess that I am uncertain if this phenomenon is frequent or infrequent; nor, indeed, is it certain why it happens. Nevertheless when it does, it is of no little importance to the angler, and being similar to the spent female, the same artificial pattern will do for both.

The Small Spurwing (*Centroptilum luteolum*) (Plate II. 19 and 20)

This fly was originally classified as one of the Pale Wateries. With only a superficial examination it is very difficult to distinguish this dun from the small Olives of the *Baëtis* genera, as it can be similar in size and colour to either the Pale Watery or Small Dark Olive. With a good lens it can readily be identified by the small spur-shaped hindwing and the single marginal intercalary veins. The eye of the male is also a brighter red than the male of the Small Dark Olive. At one time this fly used to be called the Little Sky Blue Pale Watery from the reputed colour of the wings, but from my own personal observation the wing colour of most specimens is very pale grey. Occasionally, however, the lower part of the wing is faintly white and one may say that by contrast the upper portion appears to have a bluish tinge. It has been stated by some writers that these flies are particularly attractive to trout, which will carefully select them from other species that may be simultaneously on the water. I personally have not found this to be so, but I hesitate to be dogmatic on the subject owing to the difficulty of identifying these flies positively while they are on the water. They are very common and widespread over the whole of the country, and first appear in early May, continuing through to September. Hatches are fairly prolific over most of the day in the early months of the season, but later in the year are often confined to late afternoon or early evening. The peak flight period is June. A useful point of identification for the male fly is the rather square appearance of the tail end of the body when viewed from the side. This is undoubtedly due to the thick basal portions of the male claspers. This species has been recorded as found in still water, and only recently a nymph was positively identified from Two Lakes in Hampshire.

The Little Amber Spinner (Plate VII. 89 and 92)

The popular name given to this female spinner describes it extremely well. Once it has been seen and identified it is unlikely ever to be forgotten. The upper part of the body, which appears to be faintly ringed with cream due to the segments having a cream edging, varies from a yellow-brown to a lovely pale amber colour. The under part is creamy yellow, and the last two segments are light amber. When these spinners are on the water fish seem to prefer them to many of the other spinners. At times the male spinners can also be of considerable importance to the fisherman, when the swarms along the edges of the rivers get blown on to the water in large numbers under certain conditions of wind. This often happens quite early in the evening before a fall of female spinners can be expected, and when it does the fish usually accept them readily. These male spinners also often fall spent on the water after mating. The appearance of the male which closely resembles the Large Amber Spinner is also similar to that of the male spinners of both the Iron Blue and the Small Dark Olive, but of course it is slightly larger. Also it has the same translucent white body, except that the three last segments and the thorax are of a lighter shade of brown.

The Large Spurwing (*Centroptilum pennulatum*) (Plate II. 18 and 21)

This fly used to be called the "Blue-winged Pale Watery" and I think this was an apt name for it. It is rather a pity that it is now known as the Large Spurwing. Of all the Baëtids, this is probably the easiest to identify due to its dark grey-blue wings and its very light olive-grey body, although under some conditions of light the latter appears almost cream coloured. The dun is rather localized and not particularly common, being found only in certain areas in the south of England and in parts of the north. Sometimes it is seen in the Usk Valley and in one or two isolated areas of Scotland, and appears to favour the slower flowing reaches of our rivers. It is mainly on the wing from June to August, but hatches are infrequent, depending on locality. On some occasions only the odd specimen can be seen hatching out, while on others a really prolific hatch lasting for several hours will take place. Neither I nor any angler with whom I have discussed the subject knows of any water which regularly produces good hatches throughout the fishing season. On many occasions I have seen a fish in a good lie picking these duns out in preference to other hatching duns, so there is no doubt in my mind that these are great favourites with the trout. It is similar in size to the B.W.O. An odd peculiarity of this dun is its habit of spreading its wings well apart when at rest.

The Large Amber Spinner (Plate VII. 90 and 91)

In general appearance this female spinner is similar to the Little Amber spinner, except that it is considerably larger. It has a dark, rich amber-coloured upper body, a whitish olive-coloured underbody, and conspicuous grey rings around the edges of the segments, There are several artificial patterns available for this spinner and for its smaller counterpart, although Lunn's Yellow Boy in different sizes is a good standby. The distinctive male spinner with its translucent white body, ringed with pale red or pink, also occasionally gets on the water, although not nearly so frequently as the Small Spurwing male spinner.

The Pale Evening Dun (*Procloëon pseudorufulum* (Plate I. 9)

It is surprising how little this dun is valued as an angling fly in some areas. Although not abundant it is fairly common and occurs in many localities over most of the British Isles. As it superficially resembles both the Pale Watery dun and the Small Spurwing, except that it has only two wings, it is probably mistaken for one or other of these flies by many anglers. It is a little larger than either of these duns and paler in colour. In fact it is by far the palest-coloured dun of all, with a body that is a very pale straw. As the name implies, this dun normally hatches of an evening, sometimes up to or even after dusk throughout the summer months. The sporadic hatches, when they occur, are fairly prolific. They often occur for several evenings in a row, and will then be absent for several days or even longer. In fact on some rivers it seems as if they may have quite a short season. Strange to relate, there was until comparatively recently no known artificial pattern tied for this particular species. In 1959, however, that well-known fisherman/entomologist Major Oliver Kite perfected a pattern which he called, simply, the Pale Evening Dun, and an excellent pattern it is too when this fly is on the water. This fly can be easily identified, as (apart from the much darker Pond Olive, less often seen on our rivers) it is the only river dun without hindwings. The male dun is even simpler to identify, as apart from the male Pale Watery it is the only dun with yellow eyes. This dun seems to have a preference for the slower-flowing stretches of our rivers. During the summer months, when the B.W.O. hatches are taking place regularly evening after evening, the Pale Evening dun should be remembered, as hatches of them will often occur at the same time, and on some occasions the fish will show a preference for this dun as opposed to the B.W.O.

The Pale Evening Spinner (Plate VIII. 100)

This spinner is somewhat similar to that of the Small Spurwing, but is lighter in colour, being a very pale golden brown on top and a pale greyish white underneath. Unfortunately they are of doubtful interest to the angler, being seldom seen on the water. I personally have seen only odd specimens here and there. J. R. Harris states that they are only to be seen on the wing very late in the evening, so it is possible that they fall spent after dark.

The Pond Olive (*Cloëon dipterum*) (Plate II. 28)

Although this dun is generally regarded as a still-water species, and in fact is one of the most popular and common flies of the lake or reservoir fisherman, it is not unknown on many of our slower-flowing rivers and can be found even on some of our chalk-streams in the more sluggish stretches or hatch pools. It is a medium to medium-large sized fly, very similar to the Medium Olive, except that it is much darker; it can be identified by its black-ringed tail and by the complete absence of any hindwings. Also, of course, it has only single marginal intercalary veins as opposed to the Medium Olive's double veins, which are characteristics of the *Baëtis* genus. However, it must be pointed out that these two differences apply also to the Pale Evening Dun, but as the latter has a very pale straw-coloured body, and only hatches late in the evening, there should be little difficulty in distinguishing them. Where hatches of the Pond Olive occur, the angler can use the same tactics as he would for a hatch of Medium Olives, it being extremely doubtful if the fish know the difference, although it should be mentioned the Pond Olive becomes airborne very quickly. The dun, which is widespread in all parts of the country, is seen throughout most of the summer, and the peak flight time is June and July, generally with one generation a year. However, in some years there may be two generations, with adults being on the wing during September.

The Apricot Spinner (Plate VI. 79)

This seems to be a popular name now current for the female spinner of the Pond Olive, and this is due no doubt to its distinct apricot-coloured body. The translucent wings of this spinner have strongly marked yellow-olive venation or shading along the leading edge of the wings. The spinner is of more importance than the dun to the angler, as quite often it appears that trout prefer it to other types of Olive spinners which may be on the water at the same time. I suggest that this preference by the fish may be influenced by the distinctive colour or even taste. This particular species varies a good deal in size.

The March Brown (*Rhithrogena haarupi*) (Plate III. 33 and 34)

This insect bears probably the best-known name in the fly-fisherman's world, and this is perhaps due to the fact that the artificial, which is extremely popular, can be used as a general fishing pattern on most North Country or Welsh rivers. The natural fly, in fact, is not very common, being distinctly localized; it occurs in parts of the West Country, the North of England and Scotland, but more particularly in Wales. It is a large dun with mottled fawn wings and a dull brown body. Two useful points of identification are (1) the clear patches in the middle of each forewing which are devoid of cross veins, and (2) the distinct brown streak or mark in the middle of each top (femur) leg section, this feature being a hallmark of the *Rhithrogena* genus. The March Brown is seen on the wing during late March, April or early May, and in the areas it inhabits is a very important fly to the angler. Hatches usually take place during the middle of the day in quite short yet prolific bursts. One minute not a fly can be seen on the water, the next minute the surface is literally covered with them. It prefers large brawling rivers such as the Usk, and it seems to me that the emergence is densest a little below the shallow areas of broken water. The fish will often take the hatching nymph in preference to the dun, and therefore an artificial

71. LARGE BROOK SPINNER FEMALE
Ecdyonurus torrentis

72. LARGE BROOK SPINNER MALE
Ecdyonurus torrentis

73. YELLOW MAY SPINNER FEMALE
Heptagenia sulphurea

74. YELLOW MAY SPINNER MALE
Heptagenia sulphurea

75. YELLOW UPRIGHT SPINNER FEMALE
Rhithrogena semicolorata

76. MARCH BROWN SPINNER FEMALE
Rhithrogena haarupi

77. AUTUMN SPINNER FEMALE
Ecdyonurus dispar

78. TURKEY BROWN SPINNER FEMALE
Paraleptophlebia submarginata

79. APRICOT SPINNER FEMALE
Cloeon dipterum

80. SEPIA SPINNER MALE
Leptophlebia marginata

PLATE VI. SPINNERS × 1·5

tied to represent the hatching nymph or a wet fly pattern fished just under the surface is often most effective.

The March Brown Spinner (Plate VI. 76)

This is sometimes referred to as the Great Red spinner in common with some other spinners. It is of doubtful value to the angler, as it rarely gets back on the water during the day in sufficient quantities to interest the fish. It is a large spinner, with a reddish brown body, interrupted by straw-coloured ringing.

The Olive Upright (*Rhithrogena semicolorata*) (Plate II. 25 and 26)

This is a fairly large undistinguished-looking fly, similar to but often slightly larger than a B.W.O. The body colour is grey-olive brown; it has two tails and dark blue-grey wings. At a quick glance it may also be mistaken for a large specimen of the Large Dark Olive, which it closely resembles. However, apart from the fact that it is usually a little larger it has two major points of difference, which are: (1) the hindwing is upright, and (2) in the middle of each femora, or top leg joint, there is a distinct-brown streak or mark. This, as mentioned above, is a hallmark of the *Rhithrogena* genus, occurring also in the March Brown. The duns are very common in parts of the North, Wales, Ireland and the West Country, and are usually seen on the wing from May to July. They are rare on our chalk streams, although on several occasions I have caught specimens on both the River Itchen and River Test. Hatches are often prolific, and take place usually in the late afternoon or early evening, although in some cases the duns will continue emerging after dusk. In certain localities, mostly in the tree-lined or shady areas of rivers, they also hatch during the early part of the day where the temperature of the water does not reach too high a figure.

The Yellow Upright Spinner (Plates VI. 75 and VIII. 99)

This of course is the popular name given to the spinner of the Olive Upright. When dancing in the light of the setting sun, with their bodies and wings tinted sulphury-yellow, the male spinners are exquisitely beautiful. This, combined with their habit of ascending vertically in an upright position when in flight, has resulted in their popular name. The female, which has a dull yellow-olive body, has a reputation of being disliked by the trout and so is of doubtful angling value. However, as I personally have been unable to prove or disprove this point, it might be well worth while experimenting when this spinner is observed on the water. The male spinners often swarm over the middle of the river, but are also sometimes found at a considerable distance from the water.

The Yellow May Dun (*Heptagenia sulphurea*) (Plate IV. 46 and 47)

These duns are common to most parts of the country, and their yellow bodies and pale yellow wings are a familiar sight to most anglers. It is a medium-large fly, and looks larger in flight than it actually is. They hatch out in twos and threes during the day, and are thought to be disliked by the fish. Whether or not this is due to the fact that they are unpalatable is not definitely known, but certainly I rarely see a fish take one. I think that the more likely explanation is that their emergence is more spasmodic than collective. The duns are most common in May and June. The female spinner is similar in appearance to the dun, but much paler, and the male spinner, which has a dark olive-brown body, is very similar in appearance to the male Yellow Upright spinner. An interesting feature of this species in both male and female is the colour of the eyes, which is blue-black originally in both duns and spinners, but fades with age to a pale blue. The duns sometimes hatch during the warm evenings as late as July, and some of these specimens are a very pale yellow, almost cream. These evening hatches are often more prolific than the sparse daytime hatches.

The Dusky Yellowstreak (*Heptagenia lateralis*) (Plate II. 27)

This fly is rather locally distributed. It occurs in many parts of the country, however, having a preference for smaller streams and rivers at higher altitudes. It also occurs on stony lakes where it is often quite common. In localities where it is found, hatches are often abundant and prolonged. The dun is medium-large and in appearance extremely dark. It has a dark grey-brown body and two tails. The wings are heavily veined and are a very dark grey. The dun also has fairly large upright hindwings. A useful point of identification is a distinct yellow streak on each side of the thorax just in front of the root of the forewings, and it is from this feature that its modern angling name is derived. The spinners also have this characteristic feature, and both male and female have dark olive-brown bodies. They are seen from May until September.

The Turkey Brown (*Paraleptophlebia submarginata*) (Plate III. 35)

This is an uncommon species, occurring in the medium-paced streams and rivers of South and South-west England and parts of the North of England. With its fawn-coloured mottled wings, dark brown body and three tails, it is quite a distinctive dun of a similar size or even slightly larger than the B.W.O. Its season is from mid-May to mid-June. Hatches are very sparse, and it is unusual to see more than one or two of these duns on the water at the same time. The Turkey Brown is believed to be disliked by the fish, but it seems more likely that this is due to its scarcity rather than to its taste. In any case it is of little importance as an angling fly.

The Turkey Brown Spinner (Plates VI. 78 and VIII. 101)

Like the dun, the scarcity of this spinner makes it of little importance to the fisherman, and in any case it is so similar to the Sherry spinner that there is difficulty in distinguishing one from the other. The body colour of the female usually has more brown in it than that of the Sherry, but apart from this the only sure way of identification is by checking the hindwings and the venation of the forewings. The male spinner is a very beautiful port wine colour.

The Purple Dun (*Paraleptophlebia cincta*)

This is a medium-sized dun. The body is dark brown with a purple tinge, and the wings are blackish grey. Many anglers undoubtedly mistake the Purple dun for a large Iron Blue, as superficially they are very similar, but careful examination will show that the former has three tails and large upright hindwings as opposed to the two tails and small oval spurred hindwings of the latter. The Purple dun seems to have a preference for fast flowing, highly oxygenated water, and is localized in distribution. Generally, hatches are sparse, although in some favoured localities, particularly in parts of Northern England and Ireland, they can be prolific. They are to be seen on the wing during most of the summer.

The Purple Spinner

This female spinner closely resembles that of the Turkey Brown, and although it is somewhat smaller, the only sure means of identification is by the venation of the wings. The colour of the body is brownish with a purple tinge. The male is like a large Jenny spinner, but with three tails.

The Ditch Dun (*Habrophlebia fusca*)

A rather uncommon species, usually found only in small slow-flowing streams or ditches—hence its popular name. It has a brownish olive body and rather dark grey wings. Some authorities consider it similar in appearance to a large Iron Blue, but with three tails.

From my own observations, however, I consider it is very like a Blue Winged Olive, although smaller and darker, with dull pinky-brown ventral segments. The easiest method of identifying it is by the hindwing, which has a blunt spur in the middle of the front edge (see Fig. 24). Also the forewings entirely lack the small intercalary veins which are a feature of all similar species. It has a long emergence period, lasting from May to August. The female spinner has a reddish-brown body.

The Claret Dun (*Leptophlebia vespertina*) (Plate II. 24) and

The Sepia Dun (*Leptophlebia marginata*) (Plate IV. 42 and 43)

These two species are grouped together as they are similar in habitat and appearance. They are really of greater interest to the lake fisherman, but as they occasionally frequent the slower stretches of small rivers and becks, often of an acid or peaty nature, it was thought advisable to include them here. They each have a dark brownish body, dark wings and three tails. The Claret dun is certainly the more common of the two on lakes, but on rivers they are equally scarce. They are medium-large in size, although the Sepia is usually a little larger than the Claret. Both species can be readily identified, as the Sepia has a very dark brown body with widespread tails and fawn-coloured wings with dark brown veins, whereas the body of the Claret dun is a lighter claret-brown with the forewings a very dark grey colour and pale buff-coloured hindwings. They have a short season—the Sepia appearing in early April till mid-May, and the Claret from mid-May until early July. Both are rare on our chalk streams, but on two occasions I have noticed large swarms of Claret spinners on the River Itchen. The spinners of these two species are difficult to tell apart, although the spinner of the Sepia has a distinct dark smoky patch along the top leading edge of the forewing.

The Autumn Dun (*Ecdyonurus dispar*) (Plate III. 32)

This species is again rather localized in its distribution, but is much more common on many of our rivers than the preceding species. It is a large fly with darkly veined pale fawn-coloured wings. The body is pale olive-brown, and it has two tails. It is a common fly on many of our big brawling, stony rivers such as the Welsh Usk and similar rivers in the West Country. The Autumn dun is really a late summer fly, and is most populous in August and September. Hatches are rarely prolific, but the duns, emerging in steady numbers throughout most days are occasionally welcomed by the fish. Also found on the stony shores of large lakes.

The Autumn Spinner (Plate VI. 77)

This female spinner, together with the spinners of the Large Brook dun and the Late March Brown, have in past angling literature all been referred to as "Great Red Spinners". The body colour is a very dark red and the tails are exceptionally long.

The Large Brook Dun (*Ecdyonurus torrentis*) (Plate IV. 44 and 45)

It is only in recent years that this dun has been acknowledged by anglers as a separate species. In general coloration it is very similar to the March Brown or the Late March Brown, but in size is often a little larger. The dun may be recognized by the stripes of horizontal dark brown shading across the upper part of the forewings; they resemble the dark bars or markings on the side of a perch. There is, however, a pale area devoid of cross veins in the lower middle part of the forewings, the leading edges of which are often distinctly yellow, as, indeed, is the thorax. This species has a liking for the smaller streams and brooks and is seldom seen in the larger rivers where the March Brown or Late March Brown is often so prolific. It is not a particularly common fly, and is rather localized in its distribution. In common with one or two closely related species, hatches of the Large Brook dun are often sparse, although fairly continuous throughout the day. Its time is from late March until July, reaching its peak in the months of April and May.

The Large Brook Spinner (Plate VI. 71 and 72)

The spinners of this species are very similar to the spinners of the other *Ecdyonurus* species, and it is difficult to distinguish them without a careful examination. However, a point to remember is that the forewings of the Large Brook spinner have the same yellowish leading edge as the dun. The females, with their reddish-purple bodies, can often be seen during the early afternoon either singly or in small groups, laying their eggs, which they accomplish by dipping and rising over the surface. The males have exceptionally long tails and can sometimes be found swarming far distant from the water.

The Large Green Dun (*Ecdyonurus insignis*) (Plate III. 31)

This is quite a large dun with pale fawn, mottled wings, and fairly large upright hindwings. One can notice brownish diagonal bands along the sides of the dark olive-green body, which in certain lights may appear more brown than green. A distinctive feature of both dun and spinner is the pattern or markings on the underside of the body segments (see Fig. 15). A fairly common species, but rather localized, it appears in parts of the North, South Wales and the West Country. The duns hatch, usually rather sparsely, from May to September, but the best months are July and August. Although the fish occasionally feed on them, the small numbers rarely generate much interest.

The Large Green Spinner

This distinctive spinner has an olive-green body and transparent wings with heavy dark-brown veins. A characteristic of both the male and female is a smoky brown patch along the top leading edge of the forewings. According to Dr. Michael Wade, who is very familiar with this fly on the Usk, the spinners may be plentiful *over* the water, but he cannot recall any occasion when they have been plentiful on the water. Autopsies usually reveal more males than females, but very few of either. For this reason I suggest that they may be of doubtful value to fly-fishermen. Of all the male upwinged spinners it has the longest tails, and these are of such exceptional length that the Large Green male spinner can be recognized in flight solely by this characteristic. These male spinners often swarm over the water, usually the broken stretches, flying up and down some distance from the bank at a height of six to twelve inches. Mating takes place over the water at a height of between five and ten feet.

The Late March Brown (*Ecdyonurus venosus*)

In the days of Ronalds and Halford, before *R. haarupi* had been identified, both these species were known as the "March Brown". It was only comparatively recently, in 1931 to be exact, that it was realized there were two distinct species very similar in appearance. However, *E. venosus* can be readily identified by the absence of the two pale areas on the mainwings. The habitat of this species is rather localized. Hatches are usually fairly sparse and often spread out over most of the day. The dun is of very little value to the angler, as the mature nymphs have a habit of crawling up emergent vegetation or large stones or posts protruding from the water, where they hatch and immediately fly off. They are reputed to be fairly common on the Usk, but Dr. Michael Wade has said that he rarely sees them, or their spinners, so even now a certain aura of mystery still surrounds this species.

The Great Red Spinner

This is a large spinner with a distinctive mahogany-red body. It is of more interest to the angler than the dun, as, unlike the dun, it is sometimes found on the water. In common with other Red spinners of the *Ecdyonurus* genus, it makes a juicy mouthful for the trout, and as they are all very similar in general appearance, a pattern of artificial that is acceptable for this species is suitable for most of the others.

Caënis (Plate VIII. 108)

There are only a few species of these duns, but although they vary considerably in size they are somewhat similar in appearance, and to save confusion they will be treated as one fly. They appear on lakes as well as on rivers, and are widespread over the whole of the country. They are the smallest of the Upwinged flies, and some are very small indeed. Identification is fairly easy: the body is predominantly cream, it has two rather broad wings and three tails, the combination of the two latter features making it unique amongst the Upwing duns. Some species hatch during the evening, and others very early in the morning, often before 7 a.m. Fish feeding on them are difficult to catch, hence the colloquial name, Anglers' Curse. Another unique characteristic of these tiny flies is their very rapid life-cycle in the winged form. Shortly after hatching the duns may be observed on the foliage, or on one's clothing, transposing into spinners, and within a comparatively short time they will copulate; soon afterwards the females will return to the water to lay their eggs. The spinners are whiter in colour than the duns, and the tails of the males are considerably longer than those of the females. Hatches of these little flies, particularly on lakes, are often on a tremendous scale.

As previously explained, apart from the Upwinged flies listed and described in this chapter, there are at least a dozen species which have been omitted. Many of them are extremely rare, and hardly likely to be encountered by anglers; the remainder are localized (or lake flies not dealt with in this book) and it was felt that their inclusion with the more common flies already listed would render identification of the latter more difficult.

Before I close this chapter I would like to advance two ideas that I have been experimenting with. The first is a special pattern, and the second applies to a special hook for the dressing of all patterns. I should like to mention, however, that neither of these has yet had extensive testing, and I therefore only put them forward as being worthy of consideration by the reader.

(1) Certain of the Baëtid family of duns, particularly the two species Pale Watery and Small Dark Olive, seem on some days during the summer to hatch with a distinctly yellowish aura. (This very quickly fades after emergence.) So far as I have been able to ascer-

tain this is due to the upper (femora) leg joints and the underside of the thorax being bright yellow-olive as opposed to the more common pale yellow-olive or pale olive colour. Having noticed this phenomenon on certain occasions during the last two seasons, I decided to experiment with various dressings in an endeavour to simulate this yellowish appearance, as when the trout were taking these duns the normal patterns were steadfastly refused. I feel sufficiently confident in the pattern that I have evolved, and which I have decided to call the "Yellow Halo", to include the dressing for this in the appropriate chapter at the end of this book.

(2) As a result of the many photographs taken of artificial dry flies from under water, it has become obvious that the bronze hooks on which all dry flies are tied must on occasions be very apparent to the trout against the lighter background of the sky. I am therefore dressing many of my patterns now on dull silvered hooks, as in theory they should be less visible to the fish. This may not of course be a new idea, as there is little in fishing that has not been thought of or tried before, but I have not seen this mentioned in print. I should add that I had great difficulty in obtaining the type of silvered hook I required, and eventually got them from Norway. They are a round bend up eyed hook, with a medium to short shank in a galvanized finish, and are ideal, and I am now happy to inform the reader that Messers. Veniards, the fly-tying material specialists, have agreed to stock this particular hook in future.

81. LARGE DARK OLIVE SPINNER FEMALE *Baëtis rhodani*
82. LARGE DARK OLIVE SPINNER MALE *Baëtis rhodani*
83. IRON BLUE SPINNER FEMALE *Baëtis niger*

84. MEDIUM OLIVE SPINNER FEMALE *Baëtis vernus*
85. MEDIUM OLIVE SPINNER MALE *Baëtis vernus*
86. IRON BLUE SPINNER MALE *Baëtis niger*

87. PALE WATERY SPINNER FEMALE *Baëtis bioculatus*
88. PALE WATERY SPINNER MALE *Baëtis bioculatus*
89. LITTLE AMBER SPINNER FEMALE *Centroptilum luteolum*

90. LARGE AMBER SPINNER FEMALE *Centroptilum pennulatum*
91. LARGE SPURWING SPINNER MALE *Centroptilum pennulatum*
92. SMALL SPURWING SPINNER MALE *Centroptilum luteolum*

93. SHERRY SPINNER *Ephemerella ignita*
94. BLUE-WINGED OLIVE SPINNER MALE *Ephemerella ignita*
95. AUTUMN SPINNER MALE *Ecdyonurus dispar*

PLATE VII.   SPINNERS × 1

CHAPTER VI

THE FLAT-WINGED FLIES (DIPTERA)

This Order of flies is far larger than all the other Orders put together, and in all contains several thousand species. They are commonly known as the true flies. They have only two wings which lie flat along the top of the body often overlapping (except *Tipulidae* spp.). These wings are usually transparent and colourless, although in some species small areas of colour are to be seen. Some of the more common flies in this group are House-flies, Mosquitoes, Crane-flies (Daddy-long-legs), Dung-flies, Gnats, Horse-flies, Reed Smuts and various species of Chironomidae or Midges. Many are aquatic, but by far the greater number are land-bred and of little interest to the angler. The latter is concerned with only a very few of the many land-bred species, because it is only on windy or gusty days that they may be blown on to the surface of the water in sufficient numbers to arouse the fish.

The Hawthorn-fly (*Bibio marci*) (Plate V. 68), also known as St. Mark's Fly

This latter name is probably due to the fact that hatches usually start about St. Mark's Day (25th April). This is quite a large black fly with a hairy body, about 12 mm. long, an obvious characteristic of which is a pair of long, drooping hindlegs. When this fly is on the wing it can usually be identified immediately by these long trailing hind legs. It is normally found some distance from water, often over open meadowland, flying in swarms about six feet or higher above ground level. In a strong wind, numbers of insects from these large swarms are often blown on to the water. When this happens the trout take them greedily and the fly-fisherman usually has a day to remember. Unfortunately for the trout-fisher, these days are all too rare, as in most localities the season for this fly is comparatively short, lasting only about two or three weeks.

Another similar but much smaller land-bred fly of importance to the fly-fisher is the **Black Gnat**. The scientific name is often given as *Bibio johannis* (Plate II. 29), but the name Black Gnat is in fact given to a number of different species, and *B. johannis* is but one of them.

In some parts of the country *Dilophus febrilis* is the Black Gnat, in others it is *Hilara maura* or *Ocydromia glabricula*. *Bibio* spp. can be positively identified by the venation of the wings and the curious long pointed spur on the knee joints of the two front legs. (See Fig. 48.) In a similar manner *D. febrilis* can be recognized by the many-spurred projections on the knee joints of the two front legs. (See Fig. 47.) Also the males of *Hilara* can be similarly recognized by the enlarged "popeye" joint of the tibia on the forelegs. This latter species is a more important fly to the fisherman than is generally realized in areas where they abound. Dr. Michael Wade has made a study of this particular insect, and I am indebted to him for the following extract from one of his letters.

At Monkswood on the Usk in Wales, *Hilara* are the most important type of Black Gnat from the fisherman's point of view, and in May they are accepted eagerly by the trout. They can be seen flying in all directions very close to the water, and also in swarms flying up and sometimes down stream; these swarms I believe are associated with mating. They are presumably catching their prey, which consist mainly of small midges. For mating purposes these, or sometimes small seeds, are wrapped up in a silk web by the male and presented to the female; mating takes place while she unwraps this parcel. These mating couples often fall on the water and it seems to be at this time that the trout are most interested.

Apart from the above, there are many other flies which could equally come under the general heading "Black Gnats", including of course many of the *Simulium* spp. Some of these species are seen only in spring, others only in the summer or autumn, so from the angler's point of view Black Gnats may be observed at any time during the fishing season. Again, as with the Hawthorn-flies, they are usually on the water during windy conditions, but owing to their longer season, they are there more frequently.

Fig. 47. Showing spurs on the kneejoint on forelegs of *D. febrilis*.

Fig. 48. Showing long spur on knee joint on forelegs of *Bibio* Spp.

Fig. 49. Wing venation of *Bibio* Spp.

Among other land-bred species sometimes seen on the water are the following. The **Crane-Fly** or Daddy-long-legs (*Tipula* spp.). These are occasionally taken by trout when blown on to the river, but are of more importance to the lake fisherman. However, one other particular insect closely allied to the Crane-flies is of particular interest in certain areas—the **Gravel Bed**, *Hexatoma fuscipennis* (Plate I. 12). This is a dark brownish grey insect with black legs and two brownish heavily veined wings which looks very much like a miniature Crane-fly. It is a terrestrial insect, but as the pupa prefers damp ground, it is often found in abundance on the damp gravel or sandy banks of certain rivers. The Gravel Bed is very prolific, and when a hatch occurs they are often blown on to or swarm over water (which appears to attract them), dancing and dipping as the Silverhorns do. They are mostly to be encountered in late April, May or early June, and are great favourites with both fish and fishermen.

The Dung-flies (Cordilurida) (Plate I. 14)

This large family, containing over fifty different species, specimens of which are sometimes blown on to the water, is of doubtful value to the fisherman. Another fly which comes into this category is the Oak-fly, commonly known as the Downlooker.

The aquatic flies of the Diptera can be divided into two distinct groups, the Reed Smuts and the Midges, so let us deal with the former first. The life cycle of both these groups is described in Chapter I.

Reed Smuts or Black-flies (*Simulium* spp.)

These very tiny flies often hatch out in countless numbers and in many old angling books are referred to as the Black Curse. They are extremely small, the average size being less than an eighth of an inch. There are little more than a dozen species in this family, and some of the larger specimens closely resemble the terrestrial Black Gnats and are possibly mistaken for them by many anglers. The body is short and stout with the segments hard to define. The wings are short, broad and transparent and are carried flat on top of the body when at rest. Reed Smuts somewhat resemble miniature Houseflies. They range in colour from dark brown to black. They are found only in running water, as the larvae require water with a fair amount of movement. Despite the small size of these insects trout will often feed on them avidly to the exclusion of all else, and fish so engaged are extremely hard to catch, due partly to our difficulty in dressing an artificial small enough to deceive them. The larvae are often found at autopsy, the trout having browsed them off the weeds.

The Midges (Chironomids)

This family of aquatic flies is very large, comprising nearly 400 species, and is composed of medium-size and small flies of various colours. They are quite distinctive in appearance; the cylindrical body is longer than the wings, and in many cases ringed like that of a wasp. They have the characteristic flat transparent wings of the Diptera, and these wings are rooted well forward on the body, giving the fly a somewhat hunchbacked appearance. The legs are quite long. As far as the river angler is concerned they are of little interest because most of the river species are smaller than their lake counterparts, and it is doubtful whether they hatch out in sufficiently large quantities to be of any importance. Trout undoubtedly feed on them occasionally, particularly on the larvae, which have the appearance of small wormlike creatures usually of a reddish or greenish colour which live in the mud, sediment or weed on the bottom of the river or lake. The Chironomid larvae of some species are able to live at great depths in lakes, as they can survive in water with a very low oxygen content. For this reason, too, some can live in rivers that are heavily polluted. It is interesting to note that Halford in his *Dry Fly Entomology* (page 146) states that Chironomids are rarely seen on trout rivers in any quantity except in sections of the rivers below large towns which are slightly polluted, as it would seem that these conditions are more suited to the larvae. Nearly all our rivers suffer from some form of pollution nowadays, and therefore these species are much more common than they used to be. Several months after leaving the egg, the larva changes into a pupa, which remains on the bed of the river or lake for some days until it ascends to the surface, where it hatches into the adult winged form. In lakes the pupa remains for several minutes in the surface film before the winged insect emerges. It would seem that in rivers, however, the transition to the winged state is much quicker. The pupa is quite agile and resembles the nymphs of the Olives. The adult male of the species is easily distinguished from the female by its two long hairy antennae rising upwards from the front of its head. The number of Chironomid species is quite large, and they are an important group to the lake fisherman.

CHAPTER VII

THE SEDGE-FLIES (TRICHOPTERA)

FLIES WITH ROOF-SHAPED WINGS

The Sedge, or Caddis-flies as they are known in some areas, are a fairly large group containing nearly 200 different species. They are related to certain families of moths, but differ mainly in the structure of the wings. Each has four wings, but those of the moth are covered with tiny flattened scales, while those of the Sedge-flies are covered with tiny hairs. A brief account of the life cycle of the latter is given in Chapter I.

Like moths, many of the Sedges are nocturnal, hatching out either after dark or just as the light is beginning to fade. Fortunately for the angler, some species do hatch out during the day or early evening. To generalize, the nocturnal species of Trichoptera seem to be larger, paler and less hairy than the diurnal species. Several are well known to fly-fishermen, and probably the better known of all of these daytime Sedges are the Grannom and Welshman's Button, although a Sedge which is probably the most common of all to be seen at the waterside is the Silverhorns. These small to medium-sized flies can be seen on most days in the late afternoon right through the summer, and are generally found flying in clouds just above the surface of the water. Unfortunately, although they are so prolific they seldom seem to interest the trout, and in hundreds of autopsies on trout I have seldom come across one.

Most Sedges are of a brown shade, ranging from a pale yellow-fawn to a dark red-brown through to a dark grey-brown or almost black. The life cycle is similar to the previous group Diptera, the insects passing through three stages before emerging as adult winged flies: egg, larva and pupa. The adult fly has four wings, the front wings being slightly longer than the hindwings, and when at rest these wings cover the abdomen rather like a roof. In general structure they are similar to Ephemeroptera, though certainly not in appearance. The abdomen or body, which is tailless, is composed of nine segments, and the thorax of the usual three. They have a fairly large pair of compound eyes and a pair of antennae of varying length. Protruding from the mouth parts are two pairs of palpi called the maxillary and labial respectively (these were mentioned in Chapter I). The legs are long, slender and many-jointed. The middle joints often carry several slender spurs, and the number of these spurs on each leg can often assist the entomologist to identify the species. The Sedge in the adult winged form is unable to partake of solid food but is able to consume liquid. It is rather difficult to distinguish between the sexes, and from the angler's point of view quite unnecessary.

Despite the large numbers of species in this group, relatively few of them are of interest to the angler and of these many are common to both lake and river. The fly-fisher can imitate these insects in several ways, either as a wet fly tied to imitate the pupa as it is swimming to the surface to hatch into the adult fly, or in the winged state as it endeavours to become airborne, or again on the return of the female to lay her eggs.

In Chapter II, which deals fairly extensively with the Ephemeroptera, it was possible to give considerable detail for positive identification of the various species and families for the angler with an entomological leaning. Due to the fact that there are many more species of Sedge-flies, in a book of this nature it is just not possible to do this, and in any case, identification is so much more complex that it is not certain that even the keenest angler would want to identify precisely all the Sedges in view of the doubtful value of such knowledge.

Unfortunately, much less detailed study has been made of the Sedges than of the Ephemeroptera, possibly because nearly all the great fly-fishing writers have been habitués of the chalk streams. Consequently most of the species of significance to the angler are lumped together under the headings of "Large", "Medium" or "Small" permutated with "Red", "Brown", "Grey", etc., and artificial patterns are dressed accordingly. The fact that these general patterns are usually effective leaves the average fly-fisher disinclined to investigate the natural insect any further.

It is possible to pick out certain species that are fairly regularly seen, and without too much technical detail to describe them.

In the list on the opposite page, anglers' common names have been attributed to specific species which are identified by their scientific names. It is more than likely, however, that fishermen have used, and still use, these common names not only for those identified in the

LIST OF FISHERMAN'S SEDGE-FLIES (Trichoptera)
(In order of size)

ANGLER'S NAME	ENTOMOLOGIST'S NAME	ANTERIOR WING LENGTH (approx.)
Great Red Sedge, Murragh or Northern Bustard	Phryganea grandis—P. striata	20 to 27 mm.
The Caperer	Halesus radiatus—H. digitatus	20–23 mm.
Large Cinnamon Sedge	Potamophylax latipennis (previously Stenophylax stellatus)	18 to 19 mm.
Silver Sedge or Grey Sedge	Odontocerum albicorne	13 to 18 mm.
Brown Sedge	Anabolia nervosa	11 to 16 mm.
Cinnamon Sedge	Limnephilus lunatus (previously Limnophilus lunatus)	14 to 15 mm.
The Welshman's Button	Sericostoma personatum	12 to 15 mm.
Black Sedge	Athripsodes nigronervosus previously Leptocerus nigronervosus Silo nigricornis	11 to 13 mm.
Grey Flag	Hydropsyche instabilis—H. pellucidula	11 to 12 mm.
Medium Sedge	Goëra pilosa	10 to 12 mm.
The Sand-fly	Rhyacophila dorsalis	Size variable
Marbled Sedge	Hydropsyche contubernalis (previously Hydropsyche ornatula)	11 to 12 mm.
The Grannom	Brachycentrus subnubilus	9 to 11 mm.
Small Silver Sedge (vide J. R. Harris)	Lepidostoma hirtum	9 mm.
Brown Silverhorns	Athripsodes cinereus (previously Leptocerus cinereus)	8 to 10 mm.
Black Silverhorns	Mystacides azurea Mystacides nigra Athripsodes aterrimus (previously leptocerus aterrimus)	8 to 9 mm.
Grouse Wing or Grouse and Green	Mystacides longicornis	8 to 9 mm.
Small Red Sedge	Tinodes waeneri	8 mm.
Small Yellow Sedge	Psychomyia pusilla	5 to 6 mm.

list, but for any other species (and there are many) that superficially resemble them.

Also it should be noted that many of the Sedge-flies have now been re-named. Where this occurs in the list, the old name is given in parentheses.

One of the species in the above list, *H. contubernalis*, I consider to be sufficiently common to be included, although it has not appeared previously in fishermen's lists of Sedge-flies, and I suggest the name The Marbled Sedge.

INFORMATION ON THE VARIOUS SPECIES*

The Great Red Sedge (*Phryganea grandis* and *P. striata*)

These two species are the largest of the British Trichoptera, the length of the anterior wing varying between 20 and 27 mm. The female is generally larger than the male, as is usual with most of the Caddis-flies. The wings are mostly reddish brown mottled with lighter markings and broad with a blackish bar running along the centre. They are very widely distributed over the whole of the country, but are most common in the slower running sections of many of our large rivers such as the Clyde and the Usk. They are occasionally met on some of our chalk streams, particularly the Test. The females of this family often deposit their eggs on surface vegetation, and the egg mass can easily be recognized as it is usually in the shape of a ring. The caddis case is quite large and of a characteristic spiral shape made of small sections of plant material. They are most common in late May, June and July. The adult flies usually emerge in open water and usually hatch in the late evening.

The Caperer (*Halesus radiatus*) (Plate V. 56)

Without a doubt this is one of the best known of all the Sedge-flies, particularly in the chalk streams in the South, where the common name was derived from the habit of the female of dancing and fluttering over the water in the early evening, rising and dipping as she lays her eggs. The artificial dressing of this fly was perfected by that great river keeper and fly tier, William Lunn. This is also a large species with a broad anterior wing measuring between 20 and 23 mm. in length. It has an orange-brown body and legs, yellowish-brown mottled wings and black eyes, and in all the species I have examined a most noticeable feature has been a dark striate marking down the centre of the wings. It is widely distributed over the country and hatches during the early or late evening. It is an autumn species besides being common in late August and September. The caddis case is quite large and is usually formed of pieces of cut leaf and vegetable matter. Although the hatches are not large, they are usually steady in open water and spread over a long period. For this reason, and because of their large size, the trout often feed on them eagerly.

* It should be noted that the shape, colour and markings of the wings apply to the sedge-flies at rest.

Photograph W. J. Howes

Caënis transposing on an angler's clothing.

Photograph W. J. Howes

Various types of caddis cases.

Large Cinnamon Sedge (*Potamophylax latipennis*) (Plate V. 57)

This is so similar in appearance to the Caperer that from the fisherman's point of view they can be treated as one. In place of the dark striate marking in the centre of the wings, they have a large pale area. This species, however, hatches in greater numbers and can be observed flying up and down the margins of the river as dusk approaches.

The Grey or Silver Sedge (*Odontocerum albicorne*) (Plate V. 58)

A fairly large Sedge-fly with reddish-brown eyes, and as the name implies the overall coloration is grey. The wings are a silvery grey (often with black striate marks in the centre), and tend to take on a yellowish tinge with age. The length of the anterior wing, which is rather narrow, varies between 13 and 18 mm. and useful points of identification are the long toothed antennae (these are composed of many small segments and on each of these is a small spur resembling a tooth) and the very long and hairy maxillary palps. It is widely distributed and seems to prefer the swifter flowing rivers. Hatches are often rather sparse from early June onwards and they will often be seen during the day. The larvae make a curved case of small grains of sand.

The Brown Sedge (*Anabolia nervosa*) (Plate V. 61)

A medium-sized species which varies considerably in size. The anterior wing is between 11 and 16 mm., is rather narrow and of an overall mid-brown shade with two light-coloured areas, one in the centre and one in the centre of the leading edge; these two areas often join. It is very widely distributed, and on some rivers it appears in great clouds during the evening, when both small and large specimens often appear in the same swarm or locality. During the day, however, it shelters in the bankside vegetation. The adult fly usually emerges via convenient weed beds and seldom in open water. It is an autumn species, although odd specimens will be seen as early as June. The caddis case is constructed of small grains of sand and small sticks, and it usually includes one stick two or three times as long as itself. It seems quite likely that the purpose of this long stick is to prevent predators such as fish or birds from swallowing the case, and it is noticeable that while most other caddis larvae seek protection from their enemies by camouflage or by sheltering among stones or vegetation, the Brown Sedge, in common with some other species, seeks salvation by attaching an inedible stick to the case. Another very common species of Sedge very similar in appearance, although a little smaller, is *Hydropsyche augustipennis*. It has wings of a dark golden brown colour, and the larva is of the non-case making variety.

Cinnamon Sedge (*Limnephilus lunatus*) (Plate V. 60)

A distinguished-looking species, slimly built with long, narrow anterior wings 14 to 15 mm., varying from a deep rich yellow to a cinnamon-brown with blackish markings. These wings also have several pale areas in the centre and a distinct pale crescent on the tip or trailing edge. The abdomen of the male often has a distinct greenish tinge. A very widely distributed species which first appears in June and is with us right through the summer until the early autumn. Hatches are usually rather sparse, but to the fisherman this is a useful Sedge-fly, being very common and often hatching throughout the day or early evening close to emergent vegetation in open water. The caddis larvae often construct their cases of small pieces of plant or rush stems.

A recent paper by Novak and Sehnal of the Czechoslovak Academy of Sciences describes how some species of the genus *Limnephilus*, and perhaps of other genera, exist on the Continent in their winged state from April, May or June, when they emerge from the pupa, throughout the summer in a sexually immature state. They then lay their eggs and die. During this period of diapause they appear to live in obscurity some distance from the water, to which they return only to lay their eggs. Consequently they are not often seen in summer, and because of it, it was assumed that there were two generations each year, the first when they were observed hatching from the pupa, and the second when some months later they were seen ovipositing. It is now known that they are the same generation. This destroys the popular belief that all Sedge-flies live only a short time after attaining their winged state.

Welshman's Button (*Sericostoma personatum*) (Plate III. 37)

Few fly-fishermen can have failed to have heard of this well-known Sedge-fly. There seems little doubt that originally this was the common name for a small beetle, but for some inexplicable reason that great fly-fisherman/entomologist Halford chose to apply the name to this species of Sedge-fly, and since his day this has been generally accepted. It is very common and widespread with a dark grey black body, sometimes tinged green. The length of the anterior wing varies between 12 and 15 mm. and is of a dark or golden chestnut-brown colour clothed with dense golden hairs, and sometimes with small pale patches in the rear centre of the anterior wings. The antennae are dark brown, stout and a little shorter than the wings, while the legs are golden brown. The adults often hatch in open water, sometimes in considerable numbers during the day or early evening, and the female is often to be observed carrying her large dark brown egg ball. It is usually seen in early May through to the middle of July. The larva constructs a slightly curved case of sand grains or fine gravel, with a smooth finish. It is similar to the case of the Silverhorns, although a little larger.

Black Sedge (*Athripsodes nigronervosus*)

This Sedge-fly is an all-black insect, and although similar to the Black Silverhorns it is a little larger and can quickly be distinguished from the latter as the antennae are black and scarcely annulated. The antennae of the Silverhorn are black but annulated with white. These are long and slender in both species, often up to three times the length of the wings. The anterior wings of the Black Sedge are slim, between 11 and 13 mm., and the veining is very conspicuous. They are very widely distributed and prefer the larger rivers, flying very rapidly over the water. In this particular species the female is smaller than the male, with shorter antennae and wings. It is a day-flying insect emerging from shelter in the afternoon and can be observed flying in and out close to the banks in little swarms. Their season is from June to September. The caddis case is usually constructed of fine grains of sand and is often slightly curved. Another species of Black Sedge is *Silo nigricornis* (Plate III. 40), which is so similar in appearance to *A. nigronervosus* that from the fisherman's point of view they can be grouped together. In general appearance *S. nigricornis* is a little wider in the wing and also a little smaller, between 9 and 10 mm. Also the female is often brown rather than black.

Grey Flag (*Hydropsyche pellucidula* or *H. instabilis*) (Plate V. 59)

This is one of the day-flying species, often first appearing in the early morning, flying in and out of the rushes or bankside vegetation in twos or threes but seldom in swarms. It is widely distributed and abundant, and prefers the faster flowing rivers or stretches of rivers. The anterior wing is 11 to 12 mm. long, fairly broad, and the colour is grey with blackish markings. The antennae are usually about the same length as the wings, and a characteristic feature, as in all *Hydropsyche*, is a slightly elevated and blackened ridge winding spirally round the joints. The adults emerge at the surface in open water. The larvae of this family are of the non-case making variety. They build a web of silk on or between stones on the river bed, where they lie in a silk tunnel which is an extension of this web and collect minute organisms and vegetable detritus brought by the current into the web. From observation it would seem this species usually mates in flight.

Medium Sedge (*Goëra pilosa*)

A medium-sized robust and hairy species with the anterior wing, which is very broad, between 10 and 12 mm. long. The colour varies between dark yellow and greyish yellow. It is very widely distributed and abundant and is most common during the early summer in May and June. This is another very useful Sedge to the trout fisherman, as it is also one of the day-flying species. A useful point of recognition is a small roughly circular patch in the middle lower half of the anterior wing which is free of hairs. The larvae construct a case with a central tube of sand or small stones, reinforced with larger pieces of stone at the sides.

The Sand-fly (*Rhyacophila dorsalis*) (Plate III. 38)

The common name for this species of Trichoptera was originally given in Ronalds's *Fly-Fisher's Entomology*, and it is apparent that he thought very highly of it as a fisherman's fly. It is one of the earliest Sedges to appear and is on the water from April right through the summer to the autumn. It is variable in size, and I have identified specimens with a wing length of under 10 mm. and of over 15 mm. They are difficult to identify, as apart from size they also vary a great deal in colour and appearance, although the very long legs which are a characteristic of this species can, with experience, help identification. I have seen specimens with wings of a fairly uniform sandy-brown shade, or dark brown with small whitish markings, giving a speckled appearance, or again brown with blackish markings or bars. The body is usually of a mid-brown colour (although I have personally examined many specimens that have had a greenish brown body and bright green patches on the upper legs, antennae and palps), and the antennae are slender and shorter than the wings, which are rather narrow. To help identify any *Rhyacophila* species, a magnifying glass will show that the last joint of the maxillary palpi terminates in a minute pointed spur. Also, the first and second joints are very short, the second one being globular in appearance. The Sand-fly is common and abundant and hatches throughout the day, and the larvae I suspect are of the free swimming variety.

The Marbled Sedge (*Hydropsyche contubernalis*) (Plate IV. 51)

This is a colourful-looking Sedge of medium size. The anterior wing, which is fairly narrow, varies between 11 and 12 mm. Its basic colour is a brownish green, but it has blackish brown and pale brown patches which give it a marbled appearance. It has a greenish body, orange-brown legs and black eyes. The adult insects hatch out in the early evening, and I have personally observed them actually mating in flight. They are of the same family as the Grey Flag and the larvae are of the non-case making variety.

The Grannom (*Brachycentrus subnubilus*) (Plate IV. 48)

Also known as the Greentail Fly, from the green egg sac carried by the female at the extremity of her abdomen after she has been impregnated by the male. This feature and the density of the hatches makes it the most easily recognizable of all the Sedges. The females are generally larger than the males, the respective sizes being approximately 11 mm. and 9 mm. On the many rivers they inhabit they appear in the months of April or May, and occasionally in June; the main hatch occurs about noon or a little earlier, often followed by a lesser one in the late afternoon. Thus they are a daylight-hatching Sedge, and usually among the first of the Trichoptera to appear each year. The general wing coloration is a fawn-grey with dark markings and the body colour of both the male and female, after the latter has extruded her egg sac, is a mixture of smoky grey and fawn. Before the female mates—that is, soon after emerging from the pupal sheath, and therefore prior to extruding her egg sac—the eggs which completely fill her abdomen impart a green tinge to the body. The case of the young grannom larva is square sectioned but becomes cylindrical as it matures.

The Silver Sedge (*Lepidostoma hirtum*) (Plate III. 39)

This fisherman's name has been given to *L. hirtum* by J. R. Harris, although it is generally applied to *O. albicorne*. It is quite a small species of Sedge-fly with a wing length of about 9 mm., varying in colour from grey to grey-brown, with ginger-brown legs. The antennae are slightly longer than the wings and the basal joint is longer than the head, and is covered with long erect hairs. The body is usually pale brown with a tinge of green, although I have seen specimens with a distinctly green body. The female is often to be seen carrying a bright green egg ball in a similar manner to the Grannom. This fly usually hatches out in open water early in the evening, often in large numbers. Although widely distributed, it tends to be rather localized, and is most in evidence from May to August. The shape of the wings viewed from the side when the insect is at rest is somewhat distinctive, being wide at the centre with a pronounced curve. The larva builds a square sectioned case.

The Brown Silverhorn (*Athripsodes cinereus*) (Plate IV. 52)

The Silverhorns are the most common of all the Sedges. They can be observed on most warm summer days, flying very rapidly and strongly in swarms close to the water, usually in the shelter of the banks or overhanging trees or bushes. Their extremely long antennae, often up to three times the length of the wings, are a help to identification, as in flight they curve back over the body and look like a pair of long horns. These antennae have white annulations, which give them a distinctly speckled appearance. The anterior wings, which vary between 8 and 10 mm. in length, are narrow and very variable in colour, although most specimens are brown with blackish markings. It is noticeable, and useful for identification, that when the insect is at rest and viewed from the side, there is a slight hump protruding from the top of the wing at the rear end. (This is formed by a widening of the rather narrow anterior wings at this point.) The Cinnamon Sedge also has this feature but in a more modified form. It is significant that Silverhorns are rarely, if ever, found in trout during autopsies, and it may therefore safely be assumed that they are not relished as food. Consequently imitations are of little use to the fly-fisher. However, this does not apply to the hatching pupae, which are eagerly sought by the trout and form a useful addition to their diet. The larval cases, which are usually slightly curved, are constructed from fine grains of sand. They generally inhabit the slower reaches of rivers.

The Black Silverhorn (*Mystacides azurea, M. nigra* or *A. aterrimus*) (Plate IV. 50)

This is an all-black Sedge, except for the white annulations of the antennae. The anterior wings vary between 8 and 9 mm. and in certain lights appear more steel-blue than black. The eyes are a very conspicuous dark red. They belong to the same family—Leptoceridae—as the previous species, the Brown Silverhorns, to which reference should be made for further information.

The Grouse Wing (*Mystacides longicornis*) (Plate IV. 49)

Although a common lake and pond species, it has been included here as it appears from reports that it may inhabit certain rivers. The anterior wings are greyish and have three broad grey-brown transverse bands of darker colour, giving the appearance of a grouse feather—whence the common name undoubtably originated. The antennae are white with brown annulations, and are long and slender as are most species in the family Leptoceridae. For other characteristics see under Brown Silverhorns.

Small Red Sedge (*Tinodes waeneri*)

One of the smallest of the sedges of interest to the angler, this species has anterior wings which are about 8 mm. in length. They vary in colour between yellow-brown and red-brown, and are long, narrow and hairy. The antennae are stout and usually a little shorter than the wings. They are widely distributed and abundant, and usually hatch during the evening.

Small Yellow Sedge (*Psychomyia pusilla*)

The smallest of all the fisherman's Sedge-flies listed, but nevertheless a very common and widely distributed species belonging to the same family as the preceding one. The length of the anterior wing varies between 5 and 6 mm. and is a brownish yellow colour. The antennae are usually shorter than the wings, fairly stout, and are pale yellowish white with brown annulations. They can readily be identified by a conspicuous spur or costal projection midway along the leading edge of the posterior or hindwings. On rivers or streams where they are found, they often swarm in hundreds of thousands, particularly of a late evening over the middle of the river, and my observations lead me to believe the females of this species lay their eggs by dipping.

CHAPTER VIII

THE STONEFLIES (PLECOPTERA)

THE HARD-WINGED FLIES

The flies in this group are probably the most important of all to many Midland and North Country fly-fishers. The various families in this group as a general rule favour colder water for their habitat than flies in other groups. Most of them prefer fairly fast water with a predominantly stony bottom, hence their name. With the exception of a few species they are not common in our chalk streams. It is a fairly small group of flies and according to Dr. Hynes contains a little over thirty species. In the North Country the larger species in this group are often referred to as "Mayflies" in the absence of a true Ephemerid Mayfly.

The Stoneflies vary a great deal in size, the larger specimens having a wing span of nearly two inches, and the smaller barely three-quarters of an inch. The anatomical structure is similar to that of the previous group, with the usual three-segmented thorax and the ten-segmented abdomen. The legs are of fairly uniform length but somewhat stouter than in the previous group. There are two prominent antennae mounted on the top of the head, which are usually about half the length of the body and thus considerably shorter than those of most Sedges. Unlike the latter, which are tailless, all Stoneflies have two tails which vary considerably in length. Some are as long as or slightly longer than the body, while others have tails which are reduced to mere stumps and are difficult to distinguish. The wings which are hard and shiny, lie flat along the top of the body when the fly is at rest. They are often slightly convex and particularly in some of the smaller species appear to mould around the body. There are four wings, the hindwings always being broader than the forewings. In many cases, particularly in the larger species, the wings of the male fly are foreshortened or so atrophied as to be practically non-existent, making flight impossible. A brief account of the life cycle of this group is given in Chapter 1.

Stonefly nymphs or creepers, as some are called in angling circles, are somewhat similar to the nymphs of many of the Upwinged flies, but are in most cases considerably larger. The latter, however, have three tails and the former only two. This is a point worth remembering. They are a very active and robust crawling type of nymph, and some of the larger species are carnivorous. As maturity approaches, the nymph crawls (or possibly in the case of some of the smaller species swims) to dry land, usually after dark, and seeks shelter before transposing into the fully winged adult fly. Mating is said to take place on the ground and usually two or three days later the female extrudes her eggs, which adhere to the underside of her abdomen prior to deposition. Favourite resting-places for the adult Stoneflies are the bark of trees, stones, rocks, posts, the walls of old sheds or fishing huts. The Stonefly nymph, which during its aquatic existence crawls along the bed of the river, is of little interest to the fly-fisher because no artificial can simulate the crawling action of the natural, on which trout are by no means averse to feeding. The natural nymph or creeper is a very popular bait on many of the North Country rivers. The time when a Stonefly imitation really comes into its own is when the female returns to the water to lay her eggs, and what a tempting target she makes for a hungry trout as she flutters and flops about on the surface! Also with a few of the smaller species an imitation to represent the nymph (or possibly the freshly hatched adult), as it apparently swims ashore, can be used with some effect. On the R. Test this form of fishing takes place after dusk, usually during mid-September, when the harvest moon coincides with hatches of the willow or needle flies and is referred to as "moonlight dodging".

LIST OF FISHERMAN'S STONEFLIES (Plecoptera)
(In order of size)

ANGLER'S NAME	ENTOMOLOGIST'S NAME	SIZE OF FEMALE
Large Stonefly	Perla bipunctata	
	Dinocras cephalotes	18 to 24 mm.
	Perlodes microcephala	
Medium Stonefly	Diura bicaudata	12 to 14 mm.
Yellow Sally	Isoperla grammatica	9 to 13 mm.
February Red	Taeniopteryx nebulosa	9 to 11 mm.
Willow-fly	Leuctra geniculata	8 to 11 mm.
Early Brown	Protonemura meyeri	7 to 9 mm.
Small Brown	Nemoura cinerea	6 to 9 mm.
	Nemurella picteti	
Small Yellow Sally	Chloroperla torrentium	6 to 8 mm.
Needle-fly	Leuctra fusca	5 to 9 mm.
	Leuctra hippopus	

Of the many species of Stoneflies, only about twelve are sufficiently common to be of any importance to anglers, and these are listed on page 121.

In this list, anglers' common names have been attributed to specific species which are identified by their scientific names. As with the Sedge-flies, it is more likely, however, that fishermen have used and still use these common names not only for those identified in the list, but for any other species (and there are several) that superficially resemble them.

INFORMATION ON THE VARIOUS SPECIES

The Large Stonefly (*Perla bipunctata*)

This and the following species, both with wing spans often exceeding two inches, are the largest of our Stoneflies. The former is probably the more common of the two, and is found in fast-flowing rivers with rough beds, composed of large stones which are mainly free of silt or moss. As this type of river is subject to sudden spates, the nymphs probably take shelter between the stones where they are in less danger of being swept away. It is common and widespread over all the country except Southern England and East Anglia, but most common in the North. The adults of this and the following species seldom travel far from the banks of the river and are likely to be found sheltering among stones, boulders or projections along the banks. The adults are mottled brown in appearance and most in evidence in May and June. The male fly is 16 to 23 mm. and the female 18 to 24 mm.

The Large Stonefly (*Dinocras cephalotes*)

This is very similar in size and appearance to the last species. However, it is possible to distinguish between them reasonably accurately as follows: On the preceding species (*P. bipunctata*) the pronotum—which is the top of the first segment of the thorax between the head and the forewings—is pale yellow with a black border and a black line down the centre, on each side of which lies a dark patch. On *D. cepholates* the pronotum is all black. They prefer rivers with a firm bottom, or where the bed contains a number of partly buried moss-covered stones. They occur in roughly the same areas as the preceding species, except in Ireland where they are not quite so prevalent. The adults are seen on the wing in May and June and are of a mottled brown appearance.

The Large Stonefly (*Perlodes microcephala*) (Plate IX. 118, 119)

This fly, with its wing span of nearly two inches, is a very large fly indeed, though not quite as large as the two preceding species. It is reputed to be the only large Stonefly found on our chalk streams and is often present, though only in small numbers, on the faster stonier stretches. The wings of the adult fly are mottled brown; the under body varies from cream to yellow, and the legs are often tinged with yellow. On the Usk, where this fly is common, it is often a bright yellow when first hatched, but quickly darkens with age. It is often found resting in the shelter of posts, large stones or on the bark of trees or fences several yards back from the banks. The wings of the male are often atrophied or mere stubs and the insects are incapable of flight, although active on their feet. Like many of the Stoneflies, the female is a poor flier and even when disturbed seems reluctant to take to the air. The male is 13 to 18 mm. and the female 16 to 23 mm. They are quite common, and are found mainly on lowland rivers and occasionally on the stony shores of lakes usually between April and May. This Stonefly, like most of the other large species, is probably unable to partake of food, but they can and do take liquid.

The Medium Stonefly (*Diura bicaudata*)

This is a less common species and is found only at altitudes above 1,000 ft. It is not widespread but is often abundant in the areas in which it occurs, mainly in Scotland, parts of Ireland, the Lake District and Western Wales. It is found mainly on the stony shores of lakes and small mountain streams. It is slightly larger than the more common Yellow Sally, the male being 10 to 13 mm. and the female 12 to 14 mm. The winged fly appears between April and June. They are also of a mottled brown colour.

The Yellow Sally (*Isoperla grammatica*) (Plate VIII. 106)

Of all the Stoneflies this is probably the easiest to recognize due to its conspicuous yellowish green wings, yellow-brown legs and yellow body. It is a medium-sized species and its name is often confused with one of the large yellow Upwinged flies known as the Yellow May dun, which is referred to in some areas as a Yellow Sally. The male varies between 8 and 11 mm. and the female 9 and 13 mm. It generally inhabits stony rivers or streams with a sandy or gravelly bottom, and is common but not prolific in all parts of the country except East Anglia and parts of the Midlands. The adult winged fly is often found in sheltered spots along the banks and is on the wing throughout most of the summer from April to August.

The February Red (*Taeniopteryx nebulosa*) (Plate IX. 120)

T. nebulosa is a common fly in the areas in which it occurs, but it is rather localized and is confined mainly to the North of England, parts of Wales, Scotland, and the West Country. It is probably the only Stonefly which dislikes a stony environment, being more partial to slow-flowing rivers with much vegetation. It is an early season fly and is seen usually between February and April. It is medium to small, and has red-brown wings, marked with two dark bands, the last three body segments being of a reddish brown colour. The male is 7 to 9 mm. and the female 9-11 mm. The wings of the male are often short, and it is usually the female of the species that is recognized by anglers as the February Red.

Brachyptera risi (Plate IX. 120) is more common and widespread than the preceding species, and its season is longer, extending from March to July, and in some areas may also be referred to as the February Red. It is similar in size to *T. nebulosa*, but often found in small stony streams as well as slower-paced rivers.

The Willow-fly (*Leuctra geniculata*) (Plate IX. 121)

This is also a medium to small fly, the male being 7 to 9 mm. and the female 8 to 11 mm. It is very abundant and is widespread over the whole of the country except East Anglia and Ireland. It prefers the stony beds of large streams, although it is also found in the gravelly sections of deep rivers, and is one of the few stoneflies common to our chalk streams. A late season fly, it is most in evidence from August to November. It is similar to the two following species, but has one distinct feature which should assist identification. The two long antennae have a whorl of outstanding hairs round the apex of each segment, which can be clearly seen with a magnifying glass. This Stonefly, like many of the smaller species, is often found on the limbs or bark of trees, posts or fences, where it feeds on the lichen or algae. It is a very slim fly with brownish coloured wings.

The Early Brown (*Protonemura meyeri*) (Plate IX. 122)

This is a little smaller than the previous species, the male being 5 to 8 mm. and the female 7 to 9 mm. Also the wings of this fly are a greyish brown colour, and the head usually has a transverse pale bar across the top. Apart from this, however, it can hardly be mistaken as it is an early season fly, the adult first appearing in February and seldom seen later than May. It prefers fast water with a firm bed lined with moss-covered stones or boulders. It is prolific and widespread and is often found at high altitudes.

96. CLARET SPINNER FEMALE
Leptophlebia vespertina
x1

97. CLARET SPINNER MALE
Leptophlebia vespertina
x1

98. SEPIA SPINNER FEMALE
Leptophlebia marginata
x1

99. YELLOW UPRIGHT SPINNER MALE *Rhithrogena semicolorata*
x1

100. PALE EVENING SPINNER
Procloëon pseudorufulum
x1

101. TURKEY BROWN SPINNER MALE
Paraleptophlebia submarginata
x1

102. SMALL DARK OLIVE SPINNER FEMALE *Baëtis scambus*
x1.2

103. SMALL DARK OLIVE SPINNER MALE *Baëtis scambus*
x1.2

104. POND OLIVE SPINNER MALE
Cloëon dipterum
x1

105. GREEN LACEWING
Chrysopa flava
x1

106. YELLOW SALLY
Isoperla grammatica
x1

107. SMALL YELLOW SALLY
Chloroperla torrentium
x1

108. CAËNIS SPINNER FEMALE
x1.2

109. EXUVIA OF *Torrentis*
x¾

110. POND OLIVE FEMALE SHOWING TWO PARALLEL RED LINES ALONG VENTRAL SEGMENTS
x1

PLATE VIII.   SPINNERS AND STONEFLIES, ETC. x VARIOUS

The Small Browns (*Nemoura cinerea* or *Nemurella picteti*)

These flies are very similar to those of the preceding species, both in size and appearance, but should not be confused as they are a little slimmer and of a darker colour; also they are found only in slow-flowing rivers and streams with emergent vegetation and soft, stony bottoms. The adult winged flies are to be seen from February to September.

A further species which should be mentioned is *Amphinemura sulcicollis*. This is seen mainly from April to June, but is confined usually to running water and is very common on larger streams and rivers with stony beds. It is a little smaller than the above, the female being 5 to 7 mm. in length (See Plate IX. 123.)

The Small Yellow Sally (*Chloroperla torrentium*) (Plate VIII. 107)

This is quite a small fly, the male being between 5 and 7 mm. and the female 6 and 8 mm. It is widespread and common except in the Home Counties and East Anglia, and is found in any type of water with a sandy or stony bottom, although it prefers the upland or mountainous districts. It is a slim species of a yellow or yellowish brown colour, and the adult is seen during most of the summer between April and August.

The Needle-flies (*Leuctra fusca* or *L. hippopus*, etc.) (Plate IX. 124)

These two species are the smallest of our Stoneflies and are exactly alike in appearance. Although they prefer rivers or streams with stony beds, they are widespread over the whole country, including the faster stretches of many chalk streams. The male is between 5 and 8 mm. and the female between 6 and 9 mm. As their name implies, they are exceptionally narrow flies of a dark brown colour, and although similar in shape and appearance to the Willow-fly, are much smaller. *L. fusca* is mainly a late season species, the adult winged fly appearing between August and October, while *L. hippopus* is an early season species most common from February to April.

CHAPTER IX

SUNDRY INSECTS AND OTHER FAUNA

All the flies that have so far been mentioned may be classified into definite Orders, all families in each Order having a similar life cycle. In addition, however, there are a number of miscellaneous flies belonging to various other Orders so diversified that it is impossible to deal with them in this manner. In addition, anglers are concerned with small animals such as spiders and shrimps, which are not insects at all, but which, with the indulgence of the reader, I will refer to as flies or insects for the sake of simplicity. Each of these flies will therefore be dealt with separately, and where necessary a brief account of their life history will be given. Most of the insects under this heading are terrestrial, with a life cycle of little consequence to the angler, and it is only when they find their way on to or into the water that they become of significance to the fly-fisher.

It is remarkable that nearly all the insects in this category are largely ignored by fly-fishermen, and it seems probable that the chance of an exceptional fish is sometimes missed because of it. A trout that has been born and bred in a river, or a stock fish that time has made wild and wary, and which has learned from sad experience of the dangers to be met from painful encounters with a seemingly innocuous floating insect, may live for many years and grow big and stout. A fish of this type is usually very cautious and experienced in recognizing the difference between the real and the artificial of the more common flies. However, it is surprising how often, in an unguarded moment, it can be deceived with an artificial pattern, which, although of an unfamiliar nature, nevertheless represents some kind of natural insect of an unusual variety. As an example, let us assume that a big fish has taken up residence in a good lie beneath a small overhanging bush. In fact, such a position may well have been chosen because of the cover and shade it provides. In this instance we will imagine the bush is harbouring either caterpillars or the Leaf Beetle, *Melasoma populi*, in large numbers. This small reddish-coloured insect, often found in riverside trees and bushes, like the caterpillar occasionally drops or is blown off the bush into the water. The trout is well aware of this occasional juicy tit-bit, although it may be feeding primarily on hatching duns. The really observant angler who spots one of these beetles being taken, and who can present a reasonable likeness of it to the fish, may gain a well-deserved reward.

The Alder-fly (*Sialis* spp.) (Plate V. 62)

Of all the flies mentioned in this chapter, this fly is probably the best known. It is very common and widely distributed, and is to be found near all types of water, mostly in the early part of the season, about May and June. It is similar in appearance to some dark medium size Sedge-flies, having the same roof-shaped wings when at rest. The head and legs are very dark, being almost black, and the wings are hard and shiny like those of the Stoneflies. The life cycle of this member of the Megaloptera Order is partly aquatic and partly terrestrial. The adult winged female lays her eggs in masses of between 500 and 800 on the leaves of plants or sedges overhanging the water. After about ten or twelve days the young larvae emerge and fall to the bed of the river or lake, where they live in the silt and mud on the bottom. These larvae as they mature and grow are very active and are strong swimmers. They are carnivorous, often feeding on lesser larvae and other small water creatures. They live for many months in this state, after which, according to Halford, they crawl ashore to bury themselves in the ground, forming cells in which they pupate. When the adult flies are ready to emerge from the pupa they work their way up to the surface, where the final moult takes place and the fully winged flies appear. As adults, they seem to swarm mostly in the shelter of trees or bushes, and often over the water itself. Sometimes they may be blown on to the water by accident, and when this happens trout will occasionally accept them, but as this does not happen frequently it is rather surprising that it is such an important angling fly. The artificial of this fly is often fished wet, and as the time of year nearly coincides with the period of Mayfly hatches, some anglers believe that it is taken by the trout not for the Alder which it is supposed to represent, but for the hatching Mayfly nymph, or possibly even hatching Sedges.

The Lacewings Green and Brown (Order Neuroptera) (Plate VIII. 105)

These flies are very similar to each other in general shape. There the similarity ends, as the wings of the Green Lacewing are perfectly transparent and both the wings and the green body have a most beautiful iridescent sheen, which accounts for the alternative name,

Ghostwing. The Brown Lacewing, which is normally smaller than the Green, is not such a pretty fly, as the wings are not so clear due to brownish markings; the body is also brownish. Both are only of interest on the odd occasion when they are accidentally blown on to the water, and this usually occurs in the evening. I remember a certain occasion after a day on the R. Itchen near Winchester when the light had almost gone and the evening rise was over. While actually dismantling my rod, I suddenly heard, but could not see, a number of fish rising continuously. I quickly reassembled my gear, but in spite of trying various artificials my efforts were of no avail, and after half an hour of fruitless endeavour I felt an insect crawling on my face. Fortunately I was able to capture it, and I recognized a Green Lacewing. I quickly and with difficulty changed my fly to a matching artificial which I luckily had with me, and in the very brief period that remained before the rise was over, I hooked a good fish and pricked another. I am convinced that if I had identified the fly on the water a little earlier, I would have finished the day with a brace or two.

Ants (Order Hymenoptera)

These come in either of two colours, red or black, and so far as the fish and fishermen are concerned, are only of interest in the winged stage. Most of us are familiar with the flight of ants at a certain time of the year when these normally terrestrial insects having sprouted wings, practically get in one's hair for perhaps a day or two while the flight period lasts. This transition usually takes place in the latter half of the season on calm, hot days, more often than not on sultry thundery days in late July and August. It will be appreciated that as this takes place only one or two days a year, lucky is the angler who is present at the waterside on one of these days. The insects often fall on the water in large numbers, and every fish in the river seems intent on gorging as many as possible. Providing the angler has a matching artificial he is assured of a memorable day's fishing. The moral is, of course, always to have in your fly box at least one pattern of each, red and black. They may not be used for years, but they will at least be available should the rare occasion arise.

Beetles (Order Coleoptera)

This is a very large Order, which makes it impossible to pick out many individual species of particular value to the fisherman. All sorts of different beetles find their way accidentally on to the water, and as they come in all sizes, shapes and colours, it is really a ques-

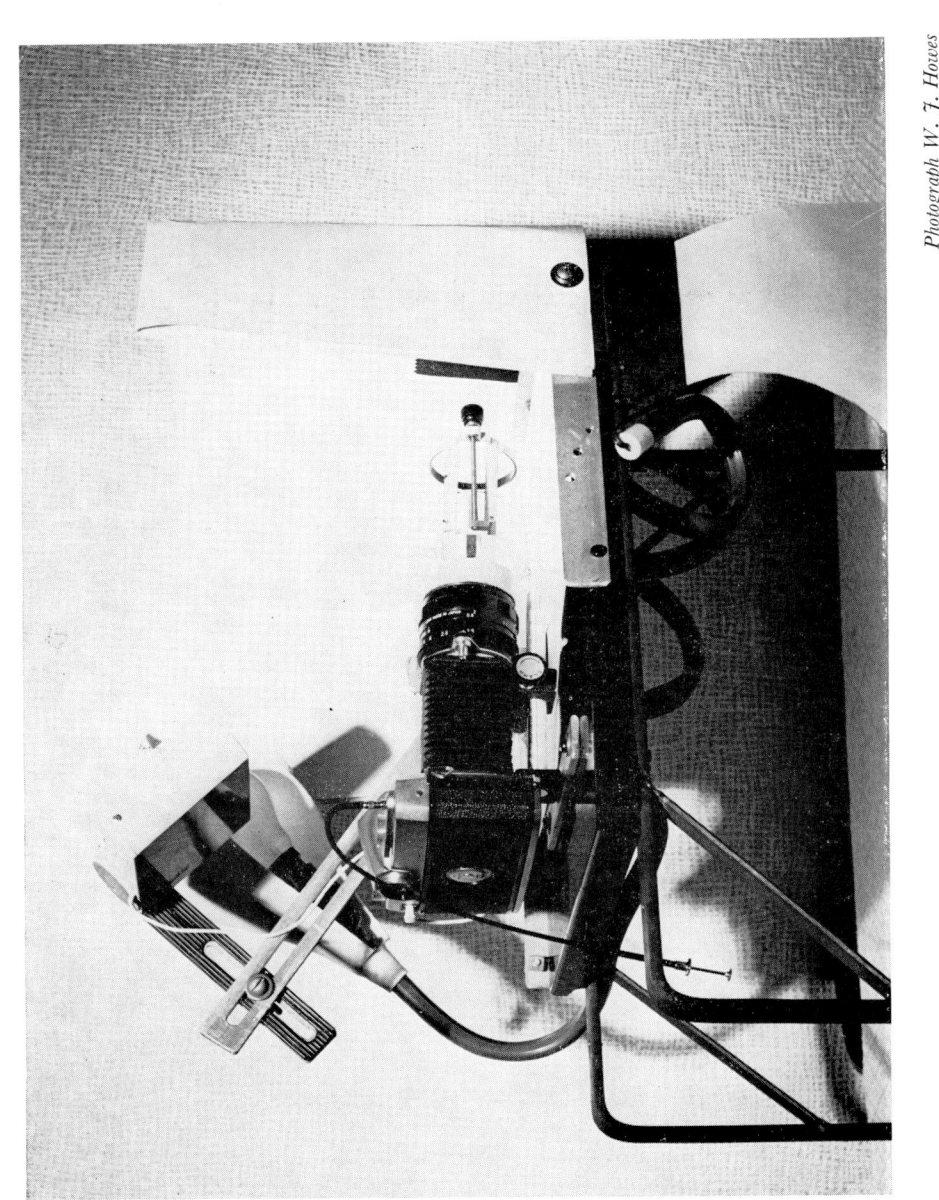

Photograph W. J. Howes

The camera with special fittings as used to take many of the colour photographs in this book.

Photograph Roy Walsh
The Upper Kennet at Kintbury—a typical chalkstream.

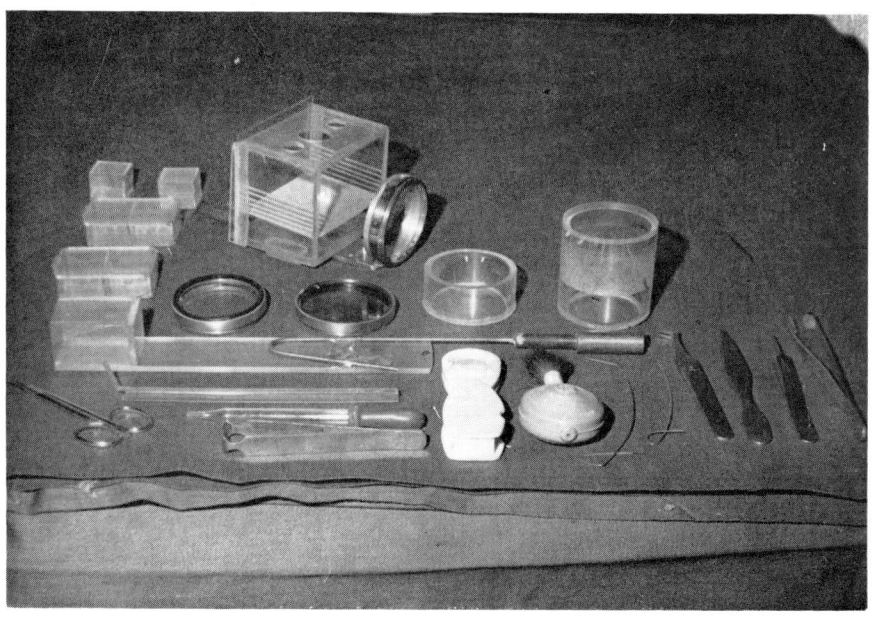

Various items of equipment necessary for photographing flies.

tion of noting whether any particular species is seen regularly on a water one fishes, and then endeavouring either to buy or tie an artificial for future use. There are, however, several types that are known and recognized generally by anglers in certain areas: the Coch-y-bonddu (*Phyllopertha horticola*). This is a beetle that appears in very large swarms in June, and is often blown on to the water. In some areas it is known as the June Bug and is most common in certain localities in Wales and Scotland. It is a small beetle about half an inch long with a metallic bluish green thorax and reddish-brown wings.

The Soldier-beetle (*Cantharis rustica*) is a very prolific species, seen in thousands during June, July and August. It is about half an inch long, quite slim, with orange-red wings which have a distinct bluish colour at the tips, and a dull yellow body. Another closely allied beetle is the Sailor-beetle (*Cantharis livida*), very similar in appearance but with dull blue wings and a more reddish body. Both of these are sometimes blown on to the water in sufficient quantities to merit the fly-fisher's attention. The common Ground-beetle (*Harpalus ruficornis*) is a beetle that rarely flies but will often fall into the river from the banks. It is of medium size, about three-quarters of an inch long and shiny black in colour. Also of course there are numerous types of water-beetles, of more interest to the fish than the fly-fisherman. However, I should mention an excellent general artificial pattern known as Eric's Beetle, invented by and named after that well-known angler Eric Horsfall-Turner. The dressing is given in Chapter XII.

Spiders (Order Araneida)

These are probably of more value to the fly-fisher than is generally realized, and a spider pattern will often tempt a trout feeding on one particular species of fly. Spiders often find their way on to the water from overhanging trees, bushes or banks, and trout generally look on them as an acceptable addition to their diet. The numbers of different species of spiders are legion, therefore it is really a question of having a few general patterns available for use when required. There is only one aquatic species in this order, *Argyroneta aquatica*.

Moths (Order Lepidoptera)

These insects are encountered on the water only occasionally, but sometimes a dour fish can be tempted with a suitable pattern, particularly at dusk. As the majority of these moths are either of a white or brown shade, it is only necessary to have a couple of patterns, one of each colour.

Grasshoppers (Order Orthoptera)

These sometimes jump on to the water from the riverside herbage, and trout undoubtedly take them, but due to the difficulty in obtaining or making satisfactory artificial patterns they are of no great account with fly-fishers generally. Grasshopper patterns are, however, sometimes employed with success by grayling fishermen.

Bees or Wasps (Order Hymenoptera)

These also are of doubtful value to anglers, although trout are not averse to feeding on any that are blown on to the water. As this seldom happens because they are strong fliers, they can more or less be discounted. I remember on one occasion, however, when a friend of mine had an exceptionally good fish on an artificial. He found a trout rising well, and after unsuccessfully trying various patterns, he moved closer to the fish the better to identify the fly it was taking. He noticed a regular procession of wasps entering and leaving a nest through a hole in the bank a little above the lie of the fish, and on those occasions when one fell on to the water and was carried struggling downstream, the fish picked it off without hesitation. Unfortunately he had no suitable imitation with him at the time, but a day or two later, when he had remedied the omission, he offered it to the fish which he netted in due course.

The Saw-fly (Order Hymenoptera) (Plate V. 63)

This insect belongs to the same Order as the bees and wasps, although it only vaguely resembles them. There are various families in this group, among them the Poplar Saw-flies, Pine Saw-flies and the Solomon's Seal Saw-flies. They are mainly dark in colour and some have touches of orange or yellow on the body. The darker variety with an almost black body is the one most commonly seen, and in Ronalds' day over a century ago, its artificial was known as the Great Dark Drone. It was probably better known then than it is today.

The Ichneumon-fly (Order Hymenoptera)

This is a most interesting fly, of which there are many types, most of which are terrestrial and of no interest to anglers. However, at least one species, *Agriotypus armatus*, is aquatic in the early stages of life. The adult flies are about half an inch long, wasp-like in appearance but somewhat slimmer, have four semi-transparent wings and longer antennae than wasps. The body and legs are black, with two or three segments in the middle of the body orange-yellow.

These flies are seldom seen on the water or even noticed by most anglers, and the reason for this is due no doubt to their curious reproduction system. After copulation, the female makes her way to the nearest water which she enters covered with a film of air in search of a host, usually a caddis larva, in which to lay her eggs. This she achieves by injecting them into the unfortunate larva's case, and when they hatch the young ichneumons feed on and gradually kill their host. As these flies are seldom actually seen on the water, but are sometimes found in trout after an autopsy, it would seem that the majority are probably taken by trout under water when the females are in the act of searching for hosts for their eggs. Ronalds, in the *Fly-Fisher's Entomology* published over a century ago, gives a dressing for this fly which in those days was apparently known as the "Orange Fly". This was tied as a floater, but in view of the foregoing remarks a wet pattern would probably be more effective.

The Freshwater Shrimp (*Gammarus pulex*) (Plate V. 65)

Crustaceans of this species are present in large numbers in most of our rivers and lakes, and form an important part of the diet of trout and grayling. As they live mainly on the bed of the river or in the weeds, they can be imitated only by a wet pattern. A fish feeding on shrimps in weed can usually be induced to accept an artificial pattern fished deep, and either a special shrimp pattern or a pheasant-tail nymph fill the bill admirably. The imitation should be cast well above, and to one side of, the weed bed in which the fish is feeding. This will give it time to sink to a level at which the fish will see it by the time it drifts down. When one judges the artificial to be about level with the fish, a slight lift may with advantage be imparted by raising the rod slightly, and more often than not the fish thinking it is a shrimp trying to escape will automatically take. This technique is termed the induced take, and is fully described in *Nymph Fishing in Practice* by Major Oliver Kite, who coined the expression.

The Water Louse (*Asellus aquaticus* or *A. meridianus*) (Plate V. 64)

Also a crustacean, it is common in many of our brooks and rivers and abundant in ponds and lakes, and like the shrimp forms an important part of a fish's diet. It is usually found in the mud or weeds on the bed of the river, where it feeds on decaying vegetable matter such as fallen leaves. As far as the angler is concerned, it can be treated in a similar manner to the shrimp, which it vaguely resembles in appearance.

CHAPTER X

NYMPHS
(EPHEMEROPTERA AND PLECOPTERA)

FOLLOWED WITH CLASSIFICATION BY TYPES

A nymph, in anglers' terms, is an aquatic fly (of a certain type) during the whole of its under-water life. Some entomologists suggest that this under-water form of life should be subdivided further into separate stages, when the insects are known as larvulae, larvae and nymphs, but in this book I am following the established custom described at the start of this paragraph. This stage is the longest of its life cycle, and can persist for weeks, months, or even years in the case of some of the large Stonefly nymphs. During this period, as they gradually grow bigger, the different species have adapted themselves variously to the conditions which appear to suit their individual physical requirements. So far as the angler is concerned, the two main Orders that are of interest and represented by this form of life, are the nymphs of the various Upwinged flies (Ephemeroptera and the Stoneflies (Plecoptera). The other two main Orders, the Diptera and the Sedges (Trichoptera), do not of course exist as true nymphs, and the equivalent form of life with these flies is known as the larval stage.

The mature nymphs of both the Upwinged flies and Stoneflies vary enormously in size from the very tiny nymphs of the Small Dark Olive (*Baëtis scambus*), which are barely a quarter of an inch long, to the very large nymphs of the Mayflies and Large Stoneflies, which are over an inch in length.

Let us first of all deal with the main difference between the nymphs of these two Orders. This is relatively simple, and with a little practice the average angler can quickly learn to spot it. The nymphs of the Upwinged flies in all cases have three distinct tails, which are quite short, being about a sixth of the length of the body. The breathing filaments or gills are grouped down each side of the abdomen. The Stonefly nymphs have only two tails, which are usually a little stouter and more prominent than the tails of the Upwing nymphs.

111. NYMPH OF E. torrentis
x1

112. NYMPH OF Ecdyonurus Spp.
x1

113. NYMPH OF Rhithrogena Spp.
x1

114. NYMPH OF L. marginata
x1

115. NYMPH OF MAYFLY
E. danica
x1

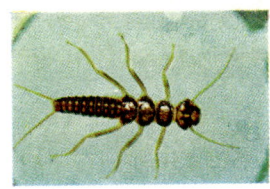

116. STONEFLY CREEPER
NYMPH OF P. microcephala
x1

117. STONEFLY CREEPER
NYMPH OF P. bipunctato
x$\frac{3}{4}$

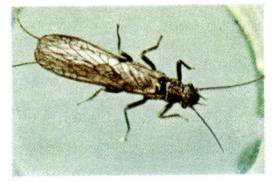

118. STONEFLY FEMALE
P. microcephala
x$\frac{3}{4}$

119. STONEFLY MALE
P. microcephala
x$\frac{3}{4}$

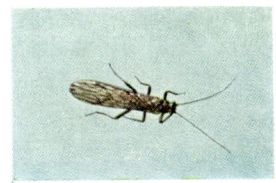

120. STONEFLY FEBRUARY RED
x1

121. STONEFLY WILLOW FLY
x1

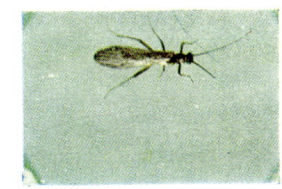

122. STONEFLY EARLY BROWN
x1

123. STONEFLY SMALL BROWN
x1

124. STONEFLY NEEDLE FLY
x1

125. SMALL STONEFLY NYMPH
x1.6

PLATE IX. NYMPHS AND STONEFLIES x VARIOUS

The gills or filaments are grouped around the thorax or shoulders of the nymph.

NYMPHS OF THE UPWINGED FLIES

In general appearance the nymphs of these flies are somewhat similar to the adult winged fly without its wings or long tails. They have a rather humped-back shape caused by the wing cases, although these are not apparent in the immature nymphs. When these wing cases first develop they are of a pale shade of brown, and as the nymphs advance in maturity they become noticeably darker. In fully mature nymphs which are at the point of emergence, the wing cases are very prominent and very dark brown. They are supposed to feed mainly on detritus and minute organisms present in the water, but whether this is in fact the case is uncertain. The various types have adapted themselves to live under widely different conditions, and this is probably why we find certain species of Upwinged flies in some localities but not in others. For instance, the nymphs of Mayflies are adapted to live in the fine silt or fine gravel on the river bottom and are therefore found only in waters that offer this type of environment. They would find it impossible to survive in the fast-flowing, stone-bedded rivers that are common in some parts of the country. Yet flat, stone-clinging nymphs of the March Brown or Autumn Dun species are very much at home in this type of river, but would find it equally impossible to live in the type of river favoured by the Mayflies. Nymphs have to absorb oxygen from the water to live. Clear pure water has a high oxygen content whereas dirty or polluted water has a low content. For this reason both trout and grayling and most nymphs are found only in our cleaner swift-flowing rivers, as they all require a fairly high oxygen content to live. All nymphs have tracheal gills along the abdomen that can, if necessary, assist the oxygenated water to flow over the body, through the walls of which the oxygen is absorbed. In the case of the various species which have adapted themselves to live in fast flowing water, these gills are not entirely essential as the water is constantly flowing past the nymph and supplying all the oxygen required. The nymphs of species that have adapted themselves to life in still or sluggish water, such as the Mayfly or Pond Olive, have overcome this problem by evolving gills (with filaments in some cases) which are capable of considerable muscular movement that provides a constant flow or circulation of water around the body, and this supplies all the oxygen required. It will doubtless be appreciated that the examples quoted here are from opposite ends of the scale, and there are many species between these two extremes that have adapted themselves to more average conditions. For this reason of course, many of the species that come in this "in between" category are widespread in their habitat and more common to many parts of the country.

Oliver Kite in his standard work *Nymph Fishing in Practice* subdivided Ephemeropteran nymphs into various groups according to their habitat, and as this seems a very sensible and practical arrangement, the same pattern will be followed in this book. The six main groups into which all the nymphs we are about to discuss can be placed are as follows:

    BOTTOM BURROWERS
    SILT CRAWLERS
    MOSS CREEPERS
    STONE CLINGERS
    LABOURED SWIMMERS
    AGILE DARTERS

The allocation of the various species to the above groups is given in detail at the end of this chapter. A brief description of the various types of nymphs comprising the above groups will now be given before we proceed to the individuals of the various species.

Bottom Burrowers (Plate IX. 115)

The only nymphs found in this country which come under this heading are the nymphs of the Mayflies. They are comparatively large and have specially shaped gills with filaments along the body. They burrow into the river bed, and live in the tunnels so formed. The head is narrow and pointed, and is provided with a large pair of mandibles which are used for excavating. The gill plates, which are very active, provide a constant flow of water past the body of the nymph, thus supplying the necessary oxygen, and perhaps this flow of water also carries minute particles of food to the nymph in its burrow.

The Silt Crawlers

The very tiny nymphs of the family Caënidae or Broadwings compose this group. So far as the river angler is concerned they are not of any great importance, and it is not therefore proposed to name the various species. They are all somewhat similar in appearance and habitat, living on the surface of the mud or silt in the slower sections of rivers or lakes. They are inactive creatures, move slowly, and are well camouflaged.

The Moss Creeper

This group embraces only two species of the genus *Ephemerella*, the nymphs of the Blue Winged Olive and the Yellow Evening dun. They are quite sturdy but rather inactive creatures which live mainly in the mosses on stones on the river bed. They also seem to favour decaying vegetation or old leaves which are trapped either in pockets on the river bed, or against obstructions such as large stones, pillars or posts in the water.

The Stone Clingers (Plate IX. 111, 112 and 113)

Nymphs in this group include several *Rhithrogena*, *Ecdyonurus* and *Heptagenia* species. The nymphs have a flattened appearance, and are wider than most of the other nymphs. They are heavily built with strong sturdy legs, specially adapted for clinging to stones in the fast-flowing waters which form their habitat. If these nymphs are observed on stones in fast currents it will be seen that they always face the flow of water, and will continue to do so even when they move, which often necessitates moving sideways or backwards. The pressure of water flowing on to the sloping flat head and thorax of this species of nymph helps them to retain their position by pressing them on to the rock or stone. They seem to favour the underside of stones during strong sunlight, and they feed on the algae (which forms on the exposed sides) when the sun is off the water.

Laboured Swimmers (Plate IX. 114)

Nymphs in this group, which includes *Paraleptophlebia*, *Leptophlebia* and *Habrophlebia*, are somewhat similar in appearance and habitat to the Moss Creepers. They have gills or plates which are long and slender, although their bodies are wide and heavily built. Their legs are as sturdy as the Moss Creepers, but they can often be observed swimming rather than crawling from place to place in a slow and leisurely manner.

The Agile Darters (Plate I. 13)

The nymphs of this group, which is by far the largest, include all species of *Baëtis*, *Centroptilum*, *Cloëon* and *Procloëon*. As the name implies, they are strong and agile swimmers, slim and streamlined in appearance with long tails, the latter fringed with hairs to provide efficient propulsion. The favourite habitat of most species is undoubtedly in clumps or beds of weed and they are to be found in all depths of water, but seem to prefer the middle or top of the weed growth. Nevertheless, they will also often be found on the moss or weed debris, and if disturbed will dart away with a good turn of speed.

Large Dark Olive (*Baëtis rhodani*)

This is probably one of the most common and widespread of all the nymphs. It is found over the whole country, but is at home more in fast-flowing than in sluggish rivers. These nymphs, agile darters, can be found clinging to the moss on stones or gravel in the faster rivers, or in the slower waters in fair quantities in all types of weed. The adult fly first appears in numbers in mid-March, or occasionally a little earlier on some waters. There are two main emergences a year. The nymphs of the spring hatching mature, for the most part, in March or April, but in some parts of the country the adult flies continue to emerge until as late as June or July. By that time, nymphs of the second emergence are already developing. Secondary hatching nymphs may appear as early as August, but the main emergence of these adult flies takes place in October and even November. Stragglers may be seen on the water in late November or December, and even, occasionally, in the early months of the year.

Iron Blue (*Baëtis pumilus* or *niger*)

These nymphs are also agile darters and are widespread over the country. Hatches are often very dense, but it is not unusual for them to occur weeks apart in some localities. *B. pumilus* is the more common of the two, *B. niger* being confined to certain local areas favouring weedy water. *B. pumilus* is typical of small stony streams, but is found in all types of water in the lowlands as well as in mountain streams. The duns emerge in May–June and again in September–October. *B. pumilus* has two main emergences a year, but whether *B. niger* has one or two is not known for certain. There is a popular belief that the heaviest hatches always occur on cold blustery days, but this has been questioned by careful observers.

Medium Olive (*Baëtis vernus*, *B. tenax* or *B. buceratus*)

The nymph of the Medium Olive is very common in the southern part of the country and is seen in greatest abundance in our chalk streams, and although widespread is not so prolific in other parts. The adults hatch mainly in the summer and are usually first seen in mid-May, with regular hatches nearly every day during the early summer and again in September. A rather small nymph of the agile darter type, they are found mostly in marginal weed beds. These nymphs are great favourites of the chalk stream trout, and it is quite common to see a good fish lying downstream of a large weed bed waiting for the hatching nymphs to drift down to it with the current.

Small Dark Olive (*Baëtis scambus*)

This is a fairly common and widespread nymph, and while abundant in limestone and chalk streams, it is also common in mountain streams. Again an agile darter but very small. It is found mainly in weed but often in the moss or stones of the river bed. The adult fly appears from May to October, but is seen in greatest numbers in July and August.

Pale Watery (*Baëtis bioculatus*)

This nymph is fairly common in Wales and the North and very common in parts of southern England, particularly in the chalk streams, as it is reputed to have a preference for calcareous waters. It is again an agile darter and finds shelter mainly in weeds, from which the duns begin to emerge in late May. It is seen on the water from then until late October, although hatches are most prolific in June, July and August.

Yellow Evening Dun (*Ephemerella notata*)

This is a rather lethargic type of nymph of the group termed moss creepers, and lives in the mosses, stones or debris on the bottom of the river with a distinct preference for water below the more swift-flowing stretches. It is one of the less common species and is confined to localized areas. It occurs in parts of Ireland, Central Wales, N.W. England and the southern parts of Devon. Specimens were taken occasionally in the Home Counties before the second World War, but it now seems practically non-existent in these areas. The adult emerges in May and June, and hatches generally occur late in the evening.

Blue Winged Olive (*Ephemerella ignita*)

This is probably our most common species, and it occurs in abundance in nearly all parts of these islands, including Ireland and Scotland. The nymph, like the preceding species, is one of the moss creepers and lives below the faster stretches of water in the moss or debris on the bottom. It is found in small brooks as well as large rivers, in the lowlands as well as the highlands, and in many big lakes. It seems equally at home in both acid and alkaline water. The adult fly is on record as having been seen in every month of the year, but first appears *en masse* about mid-June, and in some summers hatches occur daily until October, and even later in a mild autumn on some waters. Hatches may be mainly confined to late evening, just as dusk approaches. It is interesting to note that *E. ignita* spends some months in the egg, and grows rapidly after hatching in the middle of the summer. According to Dr. Macan the time spent in the egg is about ten months; other sources of information suggest a little less than this.

Turkey Brown (*Paraleptophlebia submarginata*)

A rather uncommon species of nymph, confined to rather local areas in southern and northern England. Found mainly in slow-paced rivers and streams, the nymph is a laboured swimmer, generally frequenting the thicker weed beds or debris on the bottom. The adults appear fairly early in the year, in May and June. They hatch during the day very sparsely, and it is rare to see more than one or two on the water at any time.

The Small Spurwing (*Centroptilum luteolum*)

This nymph is one of the more common species and is widespread over the whole country, but it would seem to be scarce in most of Wales except in the Usk Valley. It is again one of the agile darters and is found most frequently in weed beds in our fast or medium-paced rivers, being particularly prolific in our limestone or chalk streams; also it is reputed to be abundant in some lakes. The adults emerge from late May to early October, but are most common in June. Hatches are usually on a lavish scale and occur frequently throughout the day.

The Large Spurwing (*Centroptilum pennulatum*)

This is also one of the agile darters, but rather uncommon and found only in local areas, mainly in the southern parts of England and in the North. It also appears in the Usk Valley and in certain of our chalk streams. It would seem to have a preference for rivers of medium pace and where it occurs on the chalk streams it favours the slower stretches. The adults first emerge in late May and are about till August. Odd specimens can often be observed, but the main hatches are often weeks apart, and last only for a few days.

Pale Evening Dun (*Procloëon pseudorufulum*)

Again an agile type of nymph which is more commonly found in our medium or slow-paced rivers, or in the eddies and slow pools of faster streams. According to J. R. Harris this species is very common in Ireland, but in Scotland, England and Wales it is rather localized. It is common on some of our chalk streams and prolific on the little River Wylye, and on the slower reaches of the Upper Avon. It is also often seen on the River Usk, many of the slower-paced Yorkshire rivers and rivers in East Anglia. The adult appears mainly late in the evening, and in certain parts of rivers where the faster water gives way to the quieter pools, duns of both this species and the B.W.O. can often be observed floating down simultaneously.

March Brown (*Rhithrogena haarupi*)

By name probably one of our best known species, but in actual fact rather local in distribution. They occur in eastern Ireland, parts of Wales, the North Country and the southern part of Devon. They are more prolific in some of the Welsh rivers, particularly the Usk. The nymphs are stone clingers and appear to have a preference for the larger stony-bottomed rivers. The adults emerge spasmodically but nevertheless in very large hatches, which occur mainly in April, although sparse hatches are often to be seen in late March and early May.

Dusky Yellowstreak (*Heptagenia lateralis*)

This species is rather localized and not particularly common. It is found in parts of North Devon, Wales, Scotland, parts of Ireland and the North Country, mainly in small highland or mountain streams, rivers with stony beds and lake shores. The nymph is one of the stone clingers. The adult winged fly emerges in early May and is in evidence until late September, and is probably most common in June.

Yellow May Dun (*Heptagenia sulphurea*)

The nymphs of this fly are stone clingers, but strangely enough do not seem to have a particular preference for the more stony rivers. They are fairly widespread over the whole of the British Isles, particularly in Ireland. In the chalk streams the nymphs are often found clinging to the thicker weed stems as well as on the small pebbles of the river bed. The adults are seen in May and June, but hatches are usually very sparse and it is seldom that more than a few flies are seen together. They are sometimes found in lakes, particularly in Ireland.

Purple Dun (*Paraleptophlebia cincta*)

This nymph has a preference for small fast-flowing or large medium-paced rivers, and is a laboured swimmer with a liking for the more abundant weed. It is not a common species, being rather localized, and hatches are usually sparse. Although recorded from southern England, it is comparatively rare. However, it is often seen in Wales and the West Country, and also parts of northern England, where sometimes hatches of a more prolific nature do occur. The adults emerge from May to August, mostly during the day.

The Ditch Dun (*Habrophlebia fusca*)

As the name implies, this nymph is found mainly in small streams, ditches or silted-up carriers with often only a slight flow of water. It is one of the laboured swimmers and appears quite at home in aquatic vegetation, which is often so dense that it almost chokes these small waters. The adults hatch from May to August and are rather sparse. They are a localized species, inhabiting southern England, South Wales, the North and the Midlands. On our chalk streams they are seldom seen, except on some of the smaller and slower carriers.

Claret Dun (*Leptophlebia vespertina*)

The nymph of this fly is not very common and is of far more interest to the lake fisherman than to the river angler. It is found in small stony streams of a peaty or acid nature, and is therefore very localized. This nymph is one of the laboured swimmers and lives in the stones or moss on the river bed. The adult appears in May and June, although more in the latter than in the former.

Sepia Dun (*Leptophlebia marginata*) (Plate IX. 114)

This species is similar to the previous one, and prefers similar surroundings. The mature nymph, which is a laboured swimmer, is often to be found in the shallow margins of rivers or lakes, where it hatches via emergent vegetation or in some cases by crawling up the bank. Although it is generally even less common than the Claret dun, it seems to favour rivers slightly more than the latter. The adult is seen mainly in April and the first few days in May.

Autumn Dun (*Ecdyonurus dispar*)

This is a large nymph of the flat stone-clinging variety. It is fairly common and seems to prefer stony rivers of medium to large size, either slow or fast, and is sometimes found on stony lake shores. It is most common in Devon, South Wales, Ireland and the North Country. In the past it has been reported in the South of England, but it is doubtful whether it now exists in this area. The winged dun first appears in July and is seen until September, but the most prolific hatches are undoubtedly in late August and September.

Olive Upright (*Rhithrogena semicolorata*)

The nymph of this species is very similar to the nymph of the March Brown. It is one of the stone clingers and prefers streams and rivers of a fairly fast and stony nature. It is a common species and much more widespread than the March Brown. It is, however, mainly confined to the western half of the country, Scotland and Ireland, although it has been recorded from the Test and the Itchen. The adult fly emerges in late April or early May and continues until the end of July. It is most common in June. Very occasionally, in a cool summer, it may be seen in August and early September. Nymphs of this species are reputed to be very sensitive to a rise in temperature, and on particularly hot days in summer are subject to a high mortality rate.

Large Brook Dun (*Ecdyonurus torrentis*) (Plate IX. 111)

One of the flat stone-clinging nymphs with a distinct preference for small stony streams or brooks and the upper reaches of stony rivers. This species is common throughout the country, with the exception of southern and eastern England, although even where found appears to be selective in its choice of location. The adults emerge in late March and are seen until early July, being most common in April and May.

Pond Olive (*Cloëon dipterum*) (Plate I. 13.)

As the name implies, this nymph is confined mainly to small ponds and lakes, but it does occur occasionally in the almost static stretches of some rivers. It is one of the agile darters and lives in the weed. The adults are seen throughout the summer, but are most common in June and July, although in some seasons adults appear again in September as the result, it is believed, of a second generation.

Dark Olive (*Baëtis atrebatinus*)

This is not a common species, particularly in this country, although according to J. R. Harris it is fairly common in Ireland. It has only been included as it sometimes occurs on some of our chalk streams and on those rivers of Yorkshire that are of a calcareous nature. The nymph is one of the agile darters and the adult is seen in small numbers in April and June, and again in September and October.

Large Green Dun (*Ecdyonurus insignis*)

One of the flat stone-clinging nymphs. This again is not a common species and is confined to local areas in the North, South Wales and the West Country. It has a preference for slightly calcareous water and is found in large, fast rivers and streams. The adults are seen from May to September but most commonly in July and August.

Late March Brown (*Ecdyonurus venosus*)

This again is one of the flat stone-clinging nymphs. It is fairly common in the areas where there are fast stony rivers, replacing the Large Brook dun as the stream broadens. The mature nymphs appear to have a preference for the edges of the rivers, no doubt due to their habit of emerging from the water via large stones or rocks projecting above the surface. The adults appear from March to June, and again in certain seasons in August and September, perhaps as a second generation.

The Mayfly (*Ephemera danica, E. vulgata*) (Plate IX. 115)

These nymphs are bottom burrowers and live in tunnels in the bed of the river. *E. danica* is fairly common and widespread over the whole of the country, including Ireland and parts of Scotland, occurring in both lakes and fast-flowing rivers and streams. The adults appear from May to July and are most prolific during late May and June. *E. vulgata*, which is less common than *E. danica*, and which is mainly confined to central England, has a preference for rivers that are slow moving and have a muddy bottom.

It is generally believed that Mayflies take two years to mature, but there is now evidence at least in the case of *E. danica* that this may not be so, and I should therefore like to quote a paragraph from Dr. T. T. Macan's scientific publication *A Key to the Nymphs of the British Species of Ephemeroptera*, page 46: "Pleskot's (1958) statement that *E. danica* completes a generation in one year has a better claim to recognition than the many which allege that development takes two years, because it is the only one supported by figures. Peart (1916, 1919) appears to be alone among British authors in believing in a one-year cycle." In the case of *E. vulgata*, however, little is definitely known.

The Angler's Curse (*Caënis spp.*)

The very tiny nymphs of the various *Caënis* species belong to the group called silt crawlers, as their main habitat is the top surface of silt and mud on the bottom of the river. An alternative name for these species is Broadwing, and although they are common to many of the slow-flowing stretches of our rivers, they are much more prolific in lakes, reservoirs, etc.

NYMPHS OF THE STONEFLIES

(NYMPHS WITH TWO TAILS ONLY)

The nymphs of the Stoneflies, known as stone creepers (Plate IX. 125), are very active creatures. Their lives are spent on the bed of the river amongst the gravel or stones. Most of these nymphs are herbivorous, but at least some of the larger species are also carnivorous, feeding on smaller nymphs of other species and on the larvae of other insects. The mature nymph, which can be readily recognized by the black wing cases, crawls ashore, usually at night, where the adult winged fly emerges. The nymphs of Stoneflies are unfortunately of little value to the fly-fisherman, but the following information is included as it may be of some help in locating the adult flies.

The life span of these nymphs is about twelve months, but species of at least two genera, namely *Perla* and *Dinocras*, have a life span of three years. However, I strongly suspect that one other species, *Perlodes microcephala*, has a span also of three years, as I personally have obtained specimens of nymphs in one locality on the same day, each in a different stage of growth. The largest of these was about 23 mm., the next about 13 mm. and the smallest barely 7 mm.; and as on this particular river hatches of this fly are limited to a maximum period of three weeks, I suspect that the two smaller specimens (of which there were many) would not be sufficiently mature for metamorphosis within the given time. Therefore it appears to me that their change into the winged state must await the following year and perhaps in the case of the smallest one the year after that.

The Large Stonefly (*Perlodes microcephala*) (Plate IX. 116)

The nymph of this species is fairly large, growing to as much as 28 mm long. It is a fairly common species and is found mainly in lowland rivers with stony beds, and occasionally on stony lake shores. On the Usk where this nymph is well known, it is often a bright yellow colour with brown markings. As I have stated, it possibly lives for two or three years under water, after which the adult winged fly emerges from March to July, depending upon locality, with the most prolific period being April and May. This is reputed to be the only large Stonefly that is common to many of our chalk streams.

The Large Stonefly (*Perla bipunctata*) (Plate IX. 117)

This and the following species are our largest Stonefly nymphs. They vary between 16 and 33 mm. It is a very abundant species, and frequents rivers and streams with very rough stony bottoms, being widespread in all parts of the country, except central, southern and eastern England. The adult appears from April to June, but is most common in May and early June.

The Large Stonefly (*Dinocras cephalotes*)

This nymph is not quite so common as the preceding species. It is similar in size and appearance, and prefers rivers with a stony or rocky bottom of a more stable nature. It occurs in the same areas with the exception of Ireland, where it is more scarce.

The Medium Stonefly (*Diura bicaudata*)

A smaller and slimmer nymph than the above large Stoneflies, it varies between 9 and 17 mm. It is common in the areas it inhabits but is rather localized and occurs only in Scotland, the Lake District, western Wales and parts of Ireland. It inhabits only stony lake shore and small stony streams on higher ground. The adult appears in April and is seen until June.

Yellow Sally (*Isoperla grammatica*)

A common and widespread species, found in all parts of the country except East Anglia and parts of the Midlands. The nymph is medium sized, between 11 and 16 mm., and occurs in stony rivers, or often in the sand or gravel of small streams, and in the North on the stony shores of lakes. The adult is seen throughout the summer from April to August.

February Red (*Taeniopteryx nebulosa*)

This is a medium to small nymph and varies between 8 and 12 mm. It is fairly common in localized parts of the North, Wales, Scotland and West Country but is scarce elsewhere. It is an early species and its winged season lasts from February to April. It is probably the only Stonefly nymph that is not found in stony parts of rivers, as it prefers sluggish water and is found in vegetation, moss or at the base of sedges or reeds.

Willow-fly (*Leuctra geniculata*)

A medium to small nymph approx. 8 to 11 mm. It is a very abundant species, and is found over most of the British Isles except East Anglia and Ireland. It is present in the stony beds of large streams and in deep gravelly-bottomed rivers, and the adult appears late in the season from August to November.

Early Brown (*Protonemura meyeri*)

This nymph, a little smaller than the preceding species, is between 8 and 10 mm. It is very abundant and prefers fast water with moss-covered stones, and is often found in rivers occupying high altitudes. The adult first appears in February and is seen until May, but is most common from March to early May.

Small Brown (*Nemoura cinerea* or *Nemurella picteti*)

The nymphs of these two species are small, between 5 and 9 mm. They are very abundant and widespread over the whole country, particularly the first named. They prefer still or slow-flowing water with emergent vegetation and mossy stones. The adults are seen from February to September.

Small Yellow Sally (*Chloroperla torrentium*)

A small nymph, varying between 7 and 9 mm. It is very abundant and widespread except for the Home Counties and East Anglia. It is found in any type of water with a stony, gravelly or sandy bottom. The adults are seen from April to August, and are most common April to June. They seem to prefer waters at high altitudes.

Needle-fly (*Leuctra fusca* or *Leuctra hippopus*)

Both these nymphs are small, the former being slightly larger, between 6½ and 9 mm. The latter is between 6 and 8 mm. They are both very common and abundant, although the former is more widespread. They prefer rivers or streams with stony bottoms. The adults of the former are most common from August to October, while those of the latter are seen mainly from February to April.

THE CLASSIFICATION OF NYMPHS BY TYPES
(COMMON SPECIES ONLY)

ENTOMOLOGIST'S NAME	ANGLER'S NAME OF FLY

Bottom Burrowers (Ephemeroptera) (Plate IX. 115)

Ephemera danica	Mayfly
,, *vulgata*	,,

Silt Crawlers (Ephemeroptera)

Caënidae—various species	Angler's Curse or Broadwing

Moss Creepers (Ephemeroptera)

Ephemerella ignita	Blue Winged Olive
,, *notata*	Yellow Evening dun.

Stone Clingers (Ephemeroptera) (Plate IX. 112 and 113)

Ecdyonurus insignis	Large Green dun
,, *dispar*	Autumn dun
,, *torrentis*	Large Brook dun
,, *venosus*	Late March Brown
Rithrogena haarupi	March Brown
,, *semicolorata*	Olive Upright
Heptagenia lateralis	Dusky Yellowstreak
,, *sulphurea*	Yellow May dun

Laboured Swimmers (Ephemeroptera) (Plate IX. 114)

Habrophlebia fusca	Ditch dun
Leptophlebia vespertina	Claret dun
,, *marginata*	Sepia dun
Paraleptophlebia submarginata	Turkey Brown
,, *cincta*	Purple dun

Agile Darters (Ephemeroptera) (Plate I. 13)

Baëtis rhodani	Large Dark Olive
,, *pumilus* or *niger*	Iron Blue
,, *vernus, tenax* or *buceratus*	Medium Olive
,, *scambus*	Small Dark Olive
,, *bioculatus*	Pale Watery
,, *atrebatinus*	Dark Olive

NYMPHS (EPHEMEROPTERA AND PLECOPTERA)

ENTOMOLOGIST'S NAME	ANGLER'S NAME OF FLY
Centroptilum luteolum	Small Spurwing
,, *pennulatum*	Large Spurwing
Procloëon pseudorufulum	Pale Evening dun
Cloëon dipterum	Pond Olive

Stone Creepers (Plecoptera) (Plate IX. 116, 117 and 125)

Perlodes microcephala	Large Stonefly
Perla bipunctata	,, ,,
Dinocras cephalotes	,, ,,
Diura bicaudata	Medium Stonefly
Isoperla grammatica	Yellow Sally
Taeniopteryx nebulosa	February Red
Leuctra geniculata	Willow-fly
Protonemura meyeri	Early Brown
Nemoura cinerea	Small Brown
Nemurella picteti	,, ,,
Chloroperla torrentium	Small Yellow Sally
Leuctra fusca	Needle-fly
Leuctra hippopus	,,

CHAPTER XI

COLLECTING AND PHOTOGRAPHY

This chapter on collecting and photography, although not strictly necessary in a book on fly identification, is included because some fishermen may well decide to take advantage of the additional knowledge which can be gained from the pursuit of these two subjects.

Collecting. This can be divided into three categories:

1. The collection of specimens for study at home when it is not convenient to do so at the waterside.
2. The collection of specimens for permanent mounted collections or preserving in formalin.
3. The collection of specimens for photography purposes.

These various aspects of collecting will be dealt with in detail later in this chapter, as it is necessary first to discuss ways and means of collecting, transporting, etc.

The Net

The most important part of the collector's equipment is the net or nets, as to do the job properly two nets are necessary. Both can be made quite easily using fine mesh curtain netting or muslin. One should be completely flat for lifting floating specimens off the surface of the water, and the other can be of the traditional bag shape for either scouring the bottom or weeds for nymphs or catching specimens flying in the air. This latter type of net is also useful for collecting specimens by sweeping through long grass, herbage or vegetation, where many insects find refuge.

The ideal net for lifting flies off the surface is one that can be fixed to the top of a rod, so as to enable the collector to capture those elusive specimens that float down beyond the reach of a normal net. This constitutes a serious problem, as such a device can easily damage a delicate rod top. However, I believe I have solved the problem and I give the details below.

A 19 inch length of brass wire, 20-gauge thickness. This is a very light fine wire but amply strong for the job. Form this into a circle

$2\frac{1}{2}$ to 3 inches in diameter at one end and solder, or better still, braze. Cover this circle with flat net. Next braze or solder a small piece of the same wire about an inch long at a very slight angle to the stem about $1\frac{1}{2}$ inches down from the net. Push the stem through the top rings, with the small 1 inch piece acting as a clamp on the top or second ring. This is all that is required, except for a piece of rubber or cork to push on to the end of the stem to prevent it dropping off the end of the rod should it become loose.

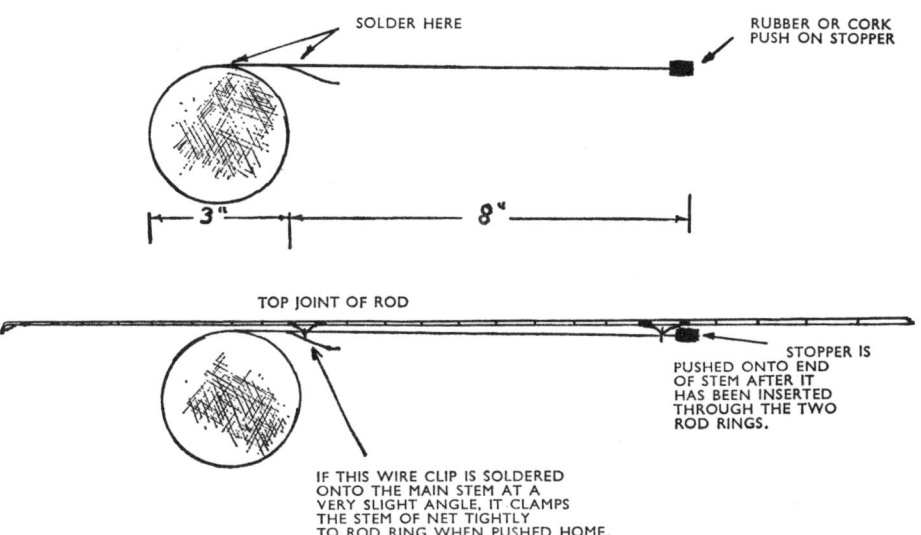

FIG. 50. Lightweight collecting net for fixing to rod top.

As will be seen from the above illustrations, this makes an extremely light and compact net which is very convenient for carrying in your bag or haversack; it can be quickly fixed or detached, and most important of all, cannot possibly damage the rod top in any way, as the only point of fixing is the clamping action on the rod ring.

Collecting Containers

Having caught your specimens, the next problem is their safe transportation. In the case of winged specimens a small glass screw-top jar with a perforated lid for ventilation is probably as good as anything, but the inside should be lined with blotting paper to absorb condensation. This is most important, as if this is not done it is most likely that your insects will reach home in a wet and bedraggled state, completely useless as specimens, through contact with the damp sides. This is particularly liable to happen on hot days.

Most Upwinged flies can easily be transferred from the net to the container by holding the wings of the specimen between forefinger and thumb, but care should be taken to hold very lightly as the wings of these flies are very delicate and easily damaged. Wet fingers incidentally are lethal. (The above method is not recommended if specimens are required eventually for transposing to spinners, as in this case, and in the case of Sedges, the wings must not be touched and the following method should be adopted.)

Captured insects can also be transferred by tapping the net over the open bottle. A mistake made by many people when collecting is to try and introduce the freshly-caught fly into the container through a partly opened lid; this is usually done to prevent specimens already in the container from escaping, but usually it results in the new arrival being injured or escaping as you try to push it through the gap between jar top and lid.

A much more efficient way is to tap the lid of the jar smartly with the palm of the hand (this knocks most of the specimens to the bottom of the container), remove the lid completely, and tap the new specimen in. If you are reasonably quick in this operation it is surprising how few flies will escape during the course of a day's collecting.

If you are prepared to take the trouble, the most efficient method of all is to construct a special collecting container. This is fairly simple to make and the construction is as follows:

Any suitable jar will suffice providing it has a good screw-on lid. Line the inside and base with blotting paper. Buy a small plastic funnel and cut off the stem, then trim off the lip until it is very slightly larger than the top of the chosen jar, so that it is a push-fit inside, with the apex and the hole down inside the jar. The resulting container will then be somewhat similar in appearance to a lobster pot and will of course work on the same principle.

Some species of flies will travel very well, while others will die quickly unless the container is kept in a cool place. When you reach home, a kitchen refrigerator is a useful place to keep specimens until they are required, but flies kept too long in this way tend either to lose their colour, or to darken considerably; this is particularly so with duns. It is as well to note that if too many specimens are collected in one container, they may damage one another. Therefore, for important specimens it is advisable to carry several small bottles, and limit one specimen to one container.

For those who wish to make an even closer study of entomology, it can be interesting to collect and hatch out nymphs. This can be a very hit-or-miss business, but providing the following procedure is carried out, a fair amount of success should be attained.

Collecting and Transporting Nymphs

Nymphs of the various species are fairly easily collected from weeds, under stones, in silt or in mud on the river bed. They can be kept in a watertight screw-top jar, but during collection the jar must be repeatedly agitated, as the nymphs will quickly die if left in completely still water. The best way to transport nymphs home after collection is in a large canvas water bucket, but even this must be repeatedly agitated.

One can with a little luck rear and hatch nymphs and larvae into winged flies under the following conditions. A small aquarium, which is the ideal container, should be treated on the exterior of three of its sides with a dark paint to keep out excess light—this is of course assuming the aquarium is to be positioned in front of a window. It should be covered on top with a hood of muslin-net curtain material, which can quite simply be attached to a wire frame. The purpose of this is two-fold: it allows plenty of necessary air to circulate in the container, and it provides a safe resting place for the flies when they hatch out.

As in this book we are concerned with flies or nymphs which require running water to survive, the container must provide it for them. This can best be provided by an ordinary low-priced air pump, which should be so installed that the air outlet discharges through a rubber or plastic tube at the bottom of the container in a way that creates a water current in a lateral direction as well as providing aeration.

The container should be set up with about an inch of gravel at the bottom and filled with water from the same stream as that from which the nymphs were taken. Also a little weed may be introduced from the same source. At this stage a word of warning: the water should be strained through a cloth before transferring it to the container, and the weed should be thoroughly washed, to prevent the introduction of aquatic creatures which might make short work of your nymphs.

If it is decided to hatch out any larvae (particularly Caddis larvae) or the larger Stonefly nymphs, they should be confined to a separate container, as many of them are carnivorous and include Ephemeropteran nymphs in their diet. With them should be included a piece of jutting rock or stone, to enable them to crawl out of the water in order to transpose. This is essential.

The nymphs themselves feed mainly on detritus and decaying vegetable matter, so do not worry about keeping your tank too clean. A fair proportion of the mature nymphs collected will usually hatch out within a matter of a week or two, but in practice it has proved very difficult to keep them alive for more than three or four weeks, so some casualties must be expected.

Let us now specifically deal with the Ephemeroptera. Nymphs which have successfully hatched into duns will, if they are allowed to remain on the net cover, within a day or two change into spinners (imagines), superb creatures of shimmering translucency. This is accomplished by the fully adult fly crawling out of its own skin which splits along the top of the thorax. With a little patience it is possible to watch this fascinating transformation take place. When handling spinners for examination, they may be picked up by the wings, and a particularly useful little tool for this delicate job is a pair of ordinary flat-bladed tweezers with a small pad of plastic foam stuck on the inside of each blade.

All these different flies, nymphs and larvae can of course be preserved indefinitely, either pinned through the body and mounted in a case, or kept in formalin for microscopic examination and reference at a later date, but the specimen should first be immersed in a 60 per cent alcohol solution for a short period before placing in the formalin. If this is not done, the specimen will not sink in the formalin properly and will quickly deteriorate.

It is not intended here to elaborate further on either of these two methods, as much has been written on them before. Also they are far more essential to the entomologist to whom microscopic examination is of prime importance. To the angler/entomologist they are not so important, as with each method the colours of the specimens quickly fade and are of little value for colour reference purposes.

It is possible to preserve an insect indefinitely by casting a block of water-clear resin around it, and do-it-yourself kits can be purchased for this purpose. The insect, after the operation is finished, is permanently embedded within the block and cannot deteriorate irrespective of the amount of handling it receives, except perhaps for a little loss of colour.

Photographing Nymphs and Flies

If one is interested in photography and can afford the necessary equipment, this is a fascinating subject of itself, offering limitless opportunities for the enthusiast. Reference collections of all the more common species can be built up in true-to-life colour and size, and if you have mastered the art of fly-tying, the advantages of having actual colour photographic models to work to in the winter months are obvious.

The problems that had to be overcome to produce the photographs that illustrate this book were many. Any reader who is thinking of trying his hand at it will find the following information of some use.

Equipment

The equipment and methods suggested here are similar to those used for the production of the photographs in this book, and for which much experimenting was necessary before success was attained.

The first essential is a good camera. The most convenient for this type of work is a 35-mm. direct-lens Reflex, which allows you to look directly through the lens and which gives very accurate focusing and positioning of your subject, most necessary in this type of work.

Film. Kodachrome II 35-mm. film is probably the best for this purpose, as this will give colour transparencies which can be studied direct or enlarged through a projecter. If the photographs are taken at 1 × magnification, this will give an actual life-size reproduction on the transparency. Kodacolour film can also be used and gives excellent colour prints, but the snag here is that in the printing the subject is considerably enlarged.

Tripod. An adjustable tripod, allowing you to tilt the camera up or down, can be used if you intend to photograph flies from various positions outdoors. This is essential for close-up work.

Electronic Flash. This is necessary for indoor photography, as it gives an exposure of one-thousandth of a second, which is desirable for these fast-moving flies.

Bellows or Extension Tubes are also useful, being far better than close-up lenses.

In the accompanying photographs (page 34), a general view of the equipment for indoor use with an electronic flash is shown, together with a view of the camera set for action (page 54). A tripod can be utilized for the latter, but a specially made platform as used is ideal. Various small perspex cells can be made and mounted in or on a sliding extension unit which is clipped or screwed on to the front of the camera. Some white cards will have to be mounted around and under each cell to throw the correct reflection of light on to the subject.

The flash unit should be set at an angle of roughly 45° overhead so that it is trained directly on to the subject. The controls of the camera should be set as follows:

Distance—infinity ∞ Aperture—*f* 16. Speed—1/25th or 1/30th of a second. The correct exposure is governed by the distance from the face of flash unit to the subject, and for a life-size reproduction the bellows should be set at magnification 1 × and for this setting the face of the flash should be approx. 7 in. from the subject to give correct exposure.

The perspex cells can be made from a 2 in. section of perspex tube with a diameter of between 2 and 3 inches, or a square tube can be made from sheet perspex. This should be sealed at each end with a disc of optically clear glass. These can be stuck on with a watertight glue, and should have a piece of perspex or glass mounted inside to form a platform on which to arrange your flies for photography. The cell should have a $\frac{5}{8}$ in. hole in the top centre for inserting and removing your specimens, and may be used in the horizontal position for photographing duns, caddis and other flies in the "at rest" position. It can also be filled with water and used for photographing nymphs in their natural environment.

The second cell is simpler as it need only be sealed with a glass disc at one end. It may be filled with water and photographed either from below, showing us what the trout sees, or from above, so that spinners, for example, can be photographed as the angler sees them.

However, let us first examine the reason for suggesting the set-up as outlined above. Many of these flies could of course be photographed in the field under natural conditions with natural daylight, but it is assumed that the average fly-fisherman will be reluctant to waste good fishing time in the field photographing specimens. Furthermore this would also necessitate transporting a great deal of equipment to the waterside.

Using electronic flash, daylight conditions can be exactly simulated, and providing specimens can be successfully transported to or bred at home, this work can then be undertaken in comfort with all the equipment at hand under ideal conditions. Also with electronic flash the exposure is constant, whereas in the field the exposure will vary greatly and one is at the mercy of the vagaries of weather.

Photographing a live insect is far from easy, as the subject more often than not refuses to keep still, and when it does finally settle at rest it is seldom in the right position. Regrettably the only answer to this problem appears to be unlimited patience, and you must be prepared to spend sometimes as much as half an hour jockeying a specimen into the right position for photographing.

We will first of all consider photographing duns, caddis, stoneflies, etc., in the horizontal cell. It has been found necessary to use a cell, as otherwise most of the time will be spent retrieving your specimen from odd corners of the room, and it is quite incredible how even in a small room a specimen can quickly vanish, perhaps never to be found again. This can be disastrous if you have only the one!

The fly should be dropped through the porthole in the top of the cell and constantly kept on the move with a steel probe. The insect eventually tires, and can then be induced to assume the right position where it is only too pleased to take a rest long enough for you to take its photograph.

We will next deal with the problem of photographing a fly from under water in the vertical cell. The only problem here is to keep the fly on the water surface while you are photographing it. Unfortunately the first reaction of a fly placed on water is to fly off, and some method had to be devised to prevent it doing so. This can be achieved as follows. Fill the chamber with water two or three days before it is required for use. During this time it will form a nearly invisible skin or film on top. The specimen should then be dropped on to this surface from a height of one to two inches; nine times out of ten the fly will land on its feet and the tension caused by the surface skin will firmly hold it in position. Sometimes, particularly with duns, they will violently flap their wings in an effort to free themselves and if you are unlucky a wing will stick in this film. When this happens the fly will have to be picked out and allowed to dry before one tries again.

Finally, with the cell in the other vertical position being photographed from above, you are now ready to take pictures of the various spinners lying spent in the surface film. It is fairly simple to prepare your subject here. Drop the spinner from an inch or two above the surface as in the previous case, and using two small steel wires, spread its wings apart and down on to the surface. Once the wings have touched the water, the fly is trapped in the right position in the surface film.

Care should be taken to ensure that the wings are pressed down evenly, as otherwise the specimen will turn on its side and is virtually impossible to right. This can only be achieved with practice, and it must be emphasized that the two expressions "patience is a virtue" or "practice makes perfect" are particularly applicable to this whole subject.

Finally, I should like to point out that many of the above problems would be simplified if the fly could be killed or doped just before photographing, but despite many months of experimenting I was unable to find any method or drug which would do this without the specimen contracting or relaxing.* It will be realized, of course, that when a fly relaxes its legs will not support the weight of its body and the wings and tails droop down, or when a fly contracts it is impossible to straighten its limbs; in either case it is then an unsuitable subject for photographing.

* In the final stages of the preparation of this book, however, I attained a certain amount of success. By using a cyanide killing bottle judiciously, some flies can be doped (not killed) and their limbs and wings positioned. This is particularly useful for photographing spinners on the water, or for quietening the very active sedges. By trial and error I have arrived at the correct time to keep various species in the cyanide bottle to reach the desired state. They are as follows:

Trichoptera	..	25 seconds
Plecoptera	..	35 seconds
Diptera ..	..	50 seconds
Ephemeroptera	..	90 seconds

CHAPTER XII

ARTIFICIALS SUGGESTED TO MATCH THE NATURAL

The lists of artificials here suggested to match the natural fly have been cut down to a minimum of two or three patterns for each, as of course it must be appreciated that the numbers of artificials available are legion. In many cases, however, there is no specific pattern to represent a certain fly, and in these cases where possible a good general pattern which is considered most suited to the particular species has been suggested. As a rough guide, against each natural fly listed will be found the hook size recommended for the matching artificial.

In this list are also included a few relatively new patterns which have consistently proved themselves in actual practice, perfected by well-known anglers in the post-war period.

THE UPWING FLIES

The Large Dark Olive. Hook size 14.
WET PATTERN   Gold-ribbed Hare's Ear—Greenwell's Glory—Water-hen Bloa.
DRY PATTERN   Rough Olive—Dogsbody—Olive Quill—Imperial.
SPINNER   Pheasant Tail Spinner—Red Spinner—Lunn's Particular.

Iron Blue Dun. Hook size 15 or 16.
WET PATTERN   Dark Watchet—Snipe and Purple.
DRY PATTERN   Iron Blue Quill—Iron Blue.
SPINNER   Houghton Ruby.

Medium Olive. Hook 15.
WET PATTERN   Greenwell's Glory—Olive Upright.
DRY PATTERN   Blue Dun—Olive Quill—Olive Dun—Dogsbody.
SPINNER   Lunn's Particular—Pheasant Tail Spinner—Port Spinner.

Small Dark Olive. Hook 15 or 16.
WET PATTERN   Poult Bloa—Snipe Bloa—Blue Upright.

DRY PATTERN   July Dun—Yellow Halo.
SPINNER   Lunn's Particular—Pheasant Tail Spinner.

Pale Watery. Hook 15.
WET PATTERN   Poult Bloa—Tup's Indispensable.
DRY PATTERN   Blue Quill—Little Marryat—Last Hope.
SPINNER   Lunn's Yellow Boy.

Yellow Evening Dun. Hook 13 or 14.
WET PATTERN
DRY PATTERN   Yellow Evening Dun.
SPINNER   Lunn's Yellow Boy—Yellow Evening Spinner.

Blue Winged Olive. Hook 14.
WET PATTERN   Gold Ribbed Hare's Ear—Poult Bloa.
DRY PATTERN   Orange Quill—B.W.O.
SPINNER   Sherry Spinner (Skues)—Port Spinner—Sherry Spinner (Woolley).

Small Spurwing. Hook 15 or 16.
WET PATTERN   Poult Bloa—Tup's Indispensable.
DRY PATTERN   Little Marryat—Last Hope.
SPINNER   Pheasant Tail Spinner—Cream Spinner (male)—Lunn's Yellow Boy.

Large Spurwing. Hook 14.
WET PATTERN   —
DRY PATTERN   Tup's Indispensable.
SPINNER   Large Spurwing Spinner—Cream Spinner (male)—Lunn's Yellow Boy.

Pale Evening Dun. Hook 14 or 15.
WET PATTERN   —
DRY PATTERN   Pale Evening Dun.
SPINNER   Yellow Evening Spinner.

Late March Brown. Hook 12 or 13.
WET PATTERN   March Brown (various).
DRY PATTERN   March Brown (various).
SPINNER   Great Red Spinner.

March Brown. Hook 12 or 13.
WET PATTERN   March Brown (various)
DRY PATTERN   March Brown (various)
SPINNER   Red Spinner.

Purple Dun. Hook 14 or 15.
WET PATTERN Snipe and Purple.
DRY PATTERN Iron Blue Quill—Purple Dun.
SPINNER   Houghton Ruby.

Claret Dun. Hook 14.
WET PATTERN —
DRY PATTERN Red Quill—Claret Dun.
SPINNER   Lunn's Particular—Claret Spinner.

Sepia Dun. Hook 14.
WET PATTERN —
DRY PATTERN Sepia Dun.
SPINNER   Pheasant Tail Spinner.

Dusky Yellowstreak. Hook 14.
WET PATTERN Dark Watchet.
DRY PATTERN Iron Blue Quill.
SPINNER   Pheasant Tail Spinner.

Autumn Dun. Hook 12 or 13.
WET PATTERN August Brown.
DRY PATTERN August Dun—March Brown.
SPINNER   Red Spinner—Great Red Spinner.

Olive Upright. Hook 14.
WET PATTERN —
DRY PATTERN Mole Fly—H.P.B. Yellow Upright.
SPINNER   Pheasant Tail spinner.

Large Brook Dun. Hook 12 or 13.
WET PATTERN March Brown.
DRY PATTERN March Brown.
SPINNER   Red Spinner—Great Red Spinner.

Pond Olive. Hook 14 or 15.
WET PATTERN Greenwell's Glory.
DRY PATTERN Olive Quill.
SPINNER   Apricot Spinner.

Dark Olive. As Large Dark Olive. Hook 14.

Large Green Dun. Hook 12 or 14.
WET PATTERN —
DRY PATTERN Large Green Dun.
SPINNER   Large Green Spinner.

Turkey Brown. Hook 14.
WET PATTERN —
DRY PATTERN   March Brown.
SPINNER   Pheasant Tail spinner.

Mayflies   Various patterns. Hook 8–10. Long Shank.
                       Special patterns: Green Drake Upright. Nevamis.

STONEFLIES

Large Stonefly	Stonefly (Veniard) Hook—Long Shank 8 or 10.
Small Brown	Partridge and Orange. Hook 14 or 15.
Medium Stonefly	February Red—Grey Duster. Hook 11–14. Stonefly (Woolley).
February Red	Grey Duster. Hook 12.
Yellow Sally	Yellow Sally. Hook 14.
Early Brown	Grey Duster. Hook 12–14.
Willow-fly.	Winter Brown—Blue Upright Wet. Hook 12–14. Grey Duster.
Needle-fly	Dark Spanish Needle. Hook 14.

SEDGE-FLIES

Cinnamon Sedge	Little Brown Sedge—Little Red Sedge—Cinnamon Sedge. Hooks 12–16.
Medium Sedge	John Storey—Medium Sedge—Brown Sedge. Hook 10 or 12.
Grey Sedge	Silver Sedge. Hook 12.
Sand-fly	Sand Fly. Hook 14.
Great Red Sedge	Brown Sedge—John Storey. Hook 8–12.
Caperer	Caperer. Hook 12–14.
Black Sedge	Black Sedge—Dark Sedge. Hook 14.
Grannom	Grannom (Henry)—Grannom (Powell). Hook 14 or 15.

OTHER FLIES

Hawthorn	Hawthorn. Hook 12.
Cow Dung	Cow-Dung Fly. Hook 12.
Oak-fly	Oak-Fly. Hook 12 to 14.

ARTIFICIALS SUGGESTED TO MATCH NATURAL

Beetles	Coch-y-bonddu—Governor—Soldier—Eric's Beetle. Hook 12–14.
Crane-fly	Daddy-Long-Legs (Hanna) — Daddy-Long-Legs (West). Hook 10 or 12.
Spiders	Black Spider—Dun Spider—Red Spider. Hook 13–15.
Blue-Bottle	Blue-bottle (Ronald's) — Blue-bottle (Hanna). Hook 12–14.
Moth	Brown Moth—Ermine Moth. Hook 10 or 12.
Angler's Curse	Last Hope—Little Marryat. Hook 17.
Caterpillar	Green Caterpillar—Arrow Fly. Hook 12 or 14.
Freshwater Shrimp	Pheasant Tail Nymph—Shrimp (Horsfall Turner). Hook 14.
Gravel Bed	Gravel Bed. Hook 13 or 14.
Black Gnat	Black Gnat (Woolley)—Black Gnat (Henry)—Black Gnat (Halford)—Black Gnat (Dunne)—Dogsbody. Hook 14–16.
Ants	Dusky Wood—Red Ant. Hook 14–17.
Alder	Alder Fly (various). Hook 12–14.
Nymphs	Pheasant Tail—P.V.C Nymph. Hook 14–17.

APPENDIX A

ARTIFICIAL FLIES

BY JOHN VENIARD

The copying of the natural insect, and improvement on patterns already established, has occupied the time of many dedicated anglers. This of course has entailed many hours spent capturing and studying the natural, a pursuit that few of us can find time for in this busy modern world.

With this book, John Goddard has removed most of this problem for us, as it is now possible to study photographs of the natural insect in the comfort of one's armchair, at any time at one's disposal.

The problem of converting steel, silk, fur and feather into a reasonable facsimile of the fly still remains, of course; but how much better it is to have a clear idea of the object to be attained, rather than trying to interpret the bald written lists of dressings found in all the fly-tying and fly-fishing manuals.

The under-water photographs of artificial flies given by John Goddard in this book show quite clearly how near one can approach the natural fly, and should be an inspiration to all fly tyers who are anxious to furnish their fly boxes with patterns which they can be confident are, at the very least, a reasonable resemblance to the object of the trout's attentions. (See Plate X.)

This does not mean that all the dressings evolved in the past must now be discarded. Many have proved to be consistently good fish takers, evolved by masters of the art after many years of experience and detailed study of the problems facing the true fly-fisherman.

Therefore, at the request of John Goddard, I have compiled a list of dressings which are associated with the natural patterns he has collected, and where possible I have named either the originator of the specimen or the person responsible for making it available for general use.

This list of dressings, together with the superb illustrations in this book, should enable us all to produce flies of a more life-like nature than has hitherto been possible.

126. LARGE DARK OLIVE MALE-DUN
126A. THE IMPERIAL (O. W. A. KITE)
127. PALE WATERY DUN

  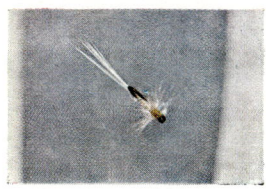

128. SEDGE-FLY THE CAPERER
128A. ARTIFICIAL SEDGE PATTERN (T. THOMAS)
127A. LAST HOPE (PALE WATERY) (J. GODDARD)

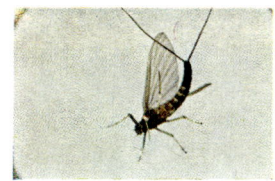  

129. MEDIUM OLIVE DUN
129A. MEDIUM OLIVE ARTIFICIAL (C. HENRY)
130. PALE WATERY SPINNER MALE

131. BLUE-WINGED OLIVE DUN
131A. B.W.O. ARTIFICIAL (D. JACQUES)
130A. PALE WATERY ARTIFICIAL SPINNER MALE

132. MEDIUM OLIVE SPINNER FEMALE
132A. LUNN'S PARTICULAR
133. VENTRAL MARKINGS ON FEMALE MAYFLY *E. danica*

PLATE X. A TROUT'S EYE VIEW
(Photographs of Naturals and matching artificials from underwater. ×1)

ARTIFICIALS LISTED IN ALPHABETICAL ORDER

(D) Dry   (S) Spinner   (W) Wet

A
Alder-Fly   (D)
Apricot Spinner (Kite)   (S)
Arrow-Fly   (D)
August Brown   (W)
August Dun   (D)

B
Black Gnat (Halford)   (D)
Black Gnat (Henry)   (D)
Black Gnat (Dunne)   (D)
Black Gnat (Woolley)   (D)
Black Sedge (Thomas)   (D)
Black Spider (Baille)
Blue-Bottle (Hanna)   (D)
Blue-Bottle (Ronald)   (D)
Blue Dun   (D)
Blue Quill   (D)
Blue Upright (Austin)   (W)
Brown Moth   (D)
Brown Sedge (Thomas)   (D)
B.W.O. (Jacques)   (D)

C
Caperer (Lunn)   (D)
Cinnamon Sedge   (D)
Claret Dun (Harris)   (D)
Claret Spinner (Harris)   (S)
Coch-y-Bonddu   (D)
Cow-Dung Fly (Hanna)   (D)
Cream Spinner (Goddard)   (S)

D
Daddy-Long-Legs (Hanna)   (D)
Daddy-Long-Legs (West)   (D)
Dark Sedge (Halford)   (D)
Dark Spanish Needle   (D)
Dark Watchet   (W)
Dogsbody   (D)
Dun Spider (Stewart)   (W)
Dusky Wood (Skues)   (D)

E
Ermine Moth (Powell)   (D)
Eric's Beetle (Horsfall-Turner)

F
February Red (Adams)   (D)

G
Gold Ribbed Hare's Ear   (W)
Governor   (D)
Grannom (Powell)   (D)
Grannom (Henry)   (D)
Gravel Bed   (D)
Great Red Spinner (Harris)   (S)
Great Red Spinner (Woolley)   (S)
Green Caterpillar   (D)
Green Drake Upright (Jacques)   (D)
Greenwell's Glory   (W)
Grey Duster   (D)

H
Hawthorn   (D)
Houghton Ruby (Lunn)   (S)

I
Imperial (Kite)   (D)
Iron Blue   (D)
Iron Blue Quill   (D)

J
John Storey   (D)
July Dun   (D)

L
Large Green Dun   (D)
Large Green Spinner   (S)
Large Spurwing Spinner (Monkhouse)   (S)
Last Hope (Goddard)
Little Brown Sedge   (D)

Little Marryat (Skues) (D)
Little Red Sedge (D)
Lunn's Yellow Boy (S)
Lunn's Particular (S)

Red Quill (D)
Red Spinner (S)
Red Spider (Stewart) (W)
Rough Olive (D)

M

March Brown (W)
March Brown (D)
Mayfly Patterns:
   Green Drake Upright (Jacques) (D)
   Nevamis (Goddard) (D)
Medium Sedge (Halford) (D)
Mole Fly (D)

S

Sandfly (Ronald) (D)
Sepia Dun (Kite) (D)
Sherry Spinner (Skues) (S)
Sherry Spinner (Woolley) (S)
Shrimp (Horsfall-Turner) (W)
Silver Sedge (D)
Snipe Bloa (W)
Snipe and Purple (W)
Soldier (Skues) (D)
Stonefly (Veniard) (D)
Stonefly (Woolley) (D)

O

Oak Fly (D)
Olive Dun (Henry) (D)
Olive Quill (D)
Olive Upright (W)
Orange Quill (D)

T

Tup's Indispensable (W)
Tup's Indispensable (D)

P

Pale Evening (Kite) (D)
Partridge and Orange (D)
Pheasant Tail,(S)
Pheasant Tail Nymph (Sawyer)
Port Spinner (Henry) (S)
Poult Bloa (W)
Purple Dun (Harris) (D)
P.V.C. Nymph (Goddard)

W

Winter Brown (D)
Water-Hen Bloa. (W)

Y

Yellow Dun (Harris) (D)
Yellow Evening Spinner (Harris) (S)
Yellow Sally (D)
Yellow Upright H.P.B. (Evans) (D)
Yellow Halo (Goddard) (D)

R

Red Ant (Lunn's) (D)

DRESSINGS LISTED IN ALPHABETICAL ORDER

Alder (Dry)

(Charles Kingsley)
   HOOK—12–14.
   TYING SILK—Crimson.
   BODY—Bronze peacock herl dyed magenta.
   WINGS—Two pairs from speckled hen wing quill, tied low over body, and with dull or outer side of feather inside.
   HACKLE—Black cock tied in front of wings, collar fashion.

(Roger Woolley)
  BODY—(1) Three strands of cock pheasant tail fibres of a good red colour.
  Rib—Waxed claret silk.
    (2) Thin bronze peacock herl, ribbed as No. 1.
    (3) Claret wool or silk ribbed with bronze peacock herl.
  WINGS—Dark brown grizzled cock hackle fibres tied low over body.
  HACKLE—Black cock tied in in front of wings.
Hackled Pattern:
  Bodies as for winged flies.
  HACKLE—A good dark blue grizzled dun cock hackle or a dark blue dun mixed with a speckled grouse hackle.
(Leonard West)
  BODY—Peacock herl.
  THORAX—Black ostrich herl.
  WINGS—Dyed grizzle blue dun cock hackle fibres.
  HORNS—Fibres from teal flank feathers.
  HACKLE—Black cock.

Apricot Spinner (Oliver Kite)
  HOOK—14.
  TYING SILK—Golden Olive.
  BODY—Swan primary herls dyed apricot.
  THORAX—Same herls doubled and redoubled.
  HACKLE—Palest honey dun.
  TAILS—Pale yellow.

Arrow-fly (Dry) (Courtney Williams)
  HOOK—12–14.
  BODY—Emerald green floss silk.
  HACKLE—White cock hackle tied palmer, i.e. from shoulder to tail.

August Brown (Wet)
  HOOK—12–14.
  TAIL—Two rabbit's whiskers.
  BODY—Light brown floss silk.
  RIB—Yellow silk.
  HACKLE—Natural brown.
  WINGS—From cock pheasant wing.

August Dun (Dry) (Roger Woolley)
  HOOK—12–14.
  TAIL—Fibres of dark ginger cock hackle.
  BODY—Brown floss silk or brown quill.
  RIB—Yellow silk.

WINGS—From mottled hen pheasant wing quill.
HACKLE—Dark ginger cock hackle.

Black Gnat (Dry) (Halford)
Male:
- HOOK—16.
- BODY—Undyed peacock quill with four close turns of black horsehair at shoulder.
- WINGS—Points of two pale blue dun cock hackles.
- HACKLE—Glossy black starling body feather.
- HEAD—Two close turns of pale maroon horsehair.

Female:
- HOOK—16.
- BODY—Stripped peacock quill dyed jet black.
- WINGS—From pale starling wing quill, tied sloping back over body.
- HACKLE—Glossy black starling body feather.

Black Gnat (Dry) (Cliff Henry)
- HOOK—16.
- TYING SILK—Black.
- BODY—Black cock hackle, flue trimmed to $\frac{1}{32}$ inch of stalk.
- WINGS—Pale starlings, tied sloping back.
- HACKLE—Black cock short in flue.

On the above pattern, David Jacques suggests that a single turn of $\frac{1}{16}$ inch wide silver tinsel be tied in at the tail end of body. This has the effect of making the body appear smaller as it separates the bend of the hook from the body of the fly.

Black Gnat (Dry) (J. W. Dunne)
- HOOK—16–17.
- TYING SILK—Olive.
- BODY—Black floss silk.
- RIB—Black tying silk.
- WINGS—A mixed bunch, consisting of about twenty fibres cut from a hackle dyed bottle-green, and about ten fibres cut from a hackle dyed fiery red-wine.
- HACKLE—Several turns of darkest honey. Wind long and then clip till fibres stand out from shank a length equal to about two-thirds of the fly. Cut a narrow V out below.

Black Gnat (Dry) (Roger Woolley)
- HOOK—16.
- BODY—Black quill.
- HACKLE—Black cock's hackle, or starling's neck feather.

Black Sedge (Dry) (Terry Thomas)

HOOK—14–10.
BODY—Black wool or chenille with black hackle tied in reverse so it slopes forward.
WING—Black moose tied on flat with the cut end to the rear.
HACKLE—Black cock.

Black Spider (Wet) (James Baillie)

HOOK—12–14.
BODY—Brown waxed silk.
HACKLE—The black/green glossy feather from the neck of a cock starling.

Blue-bottle (Dry) (Ronald's)

HOOK—12–14.
TYING SILK—brown—to be shown at head of fly.
BODY—Bright blue floss silk.
HACKLE—Short fibred black cock hackle wound down body.
WINGS—From starling wing quill.

Blue-bottle (Dry) (Hanna)

HOOK—12.
BODY—Cork covered with smoky blue balloon rubber which should be bound tightly in the middle with strong tying silk to give the fly a bulky abdomen and a slightly smaller thorax.
WINGS—Two blue dun hackle tips tied low over body.
HACKLE—Black cock tied in front of wings.
HEAD—Red cellulose varnish to simulate the large reddish head of the fly.

Blue Dun (Dry) (Roger Woolley)

HOOK—14–15.
TAIL—Dark olive cock hackle fibres.
BODY—Dark peacock quill dyed yellow, or waxed olive tying silk ribbed with gold wire.
WINGS—Dark starling wing quill or snipe wing quill.
HACKLE—Dark olive cock.
OR
TAIL—Medium blue dun cock hackle fibres.
BODY—Fur from mole or water rat spun on yellow silk.
WINGS—Dark starling or snipe.
HACKLE—Medium blue dun cock.
OR
As above but with a body formed by a well-marked stripped quill from the "eye" of a peacock tail.
This version is called the "Blue Quill".

Blue Upright (Wet) (Austin)

HOOK—14–16.
TYING SILK—Purple.
BODY—Undyed peacock's herl, stripped, taken from the side of the long stalk of the eye feather, not in the eye.
HACKLE and TAIL WHISKS—From steely blue (nearly black) cock hackle. The hackle can be wound at head only, or from shoulder to tail.

Brown Moth (Dry) (Courtney Williams)

HOOK—10–14.
TYING SILK—Orange.
BODY—Brown floss silk, rather thick.
WINGS—From a dark owl's wing feather.
HACKLE—Dark red cock hackle from shoulder to tail.

Brown Standard Sedge (Dry) (Terry Thomas)

HOOK—10–12.
BODY—Dark pheasant tail fibres wound on thick with a body hackle of ginger cock, tied in reverse so it slopes forward.
WING—Grey deer hair tied on flat with the cut end to the rear.
HACKLE—Ginger cock.

B.W.O. (Male) (Dry) (David Jacques)

HOOK—14.
TYING SILK—Orange.
BODY—Plastic (P.V.C.) sheet dyed olive (or immersed in picric acid for half hour) wound over ostrich herl dyed dirty yellow. The best results are obtained by dyeing a "grey" ostrich feather. The air in the fibres of the herl is trapped when covered by the plastic, and gives the fly buoyancy. The final colour should be that of a ripening greengage.
WINGS—Rather full, two pair from the wings of a coot.
HACKLE AND WHISKS—Light olive, particularly the stem.

Caperer (Large Sedge Dry) (William Lunn)

Winged:
HOOK—12–14.
TYING SILK—Crimson.
BODY—Four or five strands from a dark turkey tail feather, and two strands from a swan feather dyed yellow. The swan to make a ring of yellow in the centre of the body.
WINGS—From a coot's wing quill feather which has been bleached and dyed chocolate brown.
HACKLES—One medium red cock hackle and one black cock hackle, wound together in front of wings.

Hackled:
Same as for winged fly with wings omitted, and hackles longer in fibre. This fly is also known as "Welshman's Button".

Cinnamon Sedge (Dry) (Roger Woolley)
HOOK—12–14.
BODY—A strand from a cinnamon turkey tail feather.
RIB—Gold wire.
BODY HACKLE—Ginger cock. (I like this hackle to be trimmed short, tapering the shape from shoulder to tail.—Veniard.)
WINGS—Originally landrail was called for, but scarcity of this bird has resulted in substitutes such as imitation landrail, brown hen wing's quill, etc.)
SHOULDER HACKLE—Ginger cock wound in front of wings.

Halford's dressing is virtually the same except that he advocates condor herl for the body dyed cinnamon, instead of the turkey feather.

Claret Dun (Dry) (J. R. Harris)
HOOK—14.
TAIL—Dark blue dun cock hackle fibres.
BODY—Dark heron herl dyed dark claret, or mole's fur and dark claret mohair mixed on claret tying silk.
RIB—Fine gold wire.
SHOULDER HACKLE—Dark blue dun cock, six or seven turns with a V clipped out underneath.

Claret Spinner (J. R. Harris)
HOOK—14.
TAIL—Dark blue dun cock hackle fibres.
BODY—Very dark claret seal's fur spun on claret tying silk.
HACKLE—Blue dun or rusty dun cock's hackle.

Coch-y-Bonddu (Dry)
HOOK—12–14.
BODY—Two strands of bronze peacock herl.
TIP—Flat gold tinsel.
HACKLE—Coch-y-Bonddu. (A red hackle with black centre and red tips.)

Cow Dung Fly (Dry) (Hanna)
HOOK—12.
BODY—Brownish yellow chenille.
RIB—Light green tying silk.
WINGS—Tips of two darkish honey dun hackles tied on to lie flat over body.
HACKLE—Dark honey dun.

Cow Dung Fly (Dry) (Roger Woolley)
HOOK—12–13.
BODY—A mixture of yellow, dark orange, plus a little green, seal's fur.
WINGS—Landrail or substitute tied low over body.
HACKLE—Dark ginger tied in front of wings.

Cream Spinner (Male) (John Goddard)
HOOK—15.
TYING SILK—Cream.
BODY—Baby seal fur, cream.
RIB—Thin gold wire.
WINGS—Tips of two small pale blue dun cock, hackles tied "spent".
HACKLE—Cream cock hackle tied sparsely two turns only.
TAILS—Pale blue cock fibres.

Daddy-Long-Legs (Dry) (T. J. Hanna)
HOOK—10–12 long shank.
LEGS—Six fibres from a cock pheasant's tail—knotted in the middle to form joints. Four fibres tied in to point back and two to point forwards.
BODY—A thin piece of bicycle tube rubber wound up shank between the legs, to form a segmented body.
WINGS—Two brownish cock hackle tips tied "spent".
HACKLE—Natural red-brown cock hackle tied and wound behind wings.

Daddy-Long-Legs (Dry) (Leonard West)
LEGS—As for the T. J. Hanna pattern.
THORAX—Brown ostrich herl.
BODY—Fawn raffia.
RIB—Fine gold wire.
WINGS—Two red/grizzle (cree) cock hackle tips tied "spent".
HACKLE—Red/grizzle cock wound behind wings.

Ready-formed hollow plastic bodies are now available, and make realistic bodies for the "Daddy". The above dressings can be used with these bodies, using a short shanked dry fly hook.

Dark Sedge (Dry) (Halford)
HOOK—14.
BODY—Stripped condor quill dyed very dark maroon.
BODY HACKLE—Dark furnace cock.
WINGS—Landrail or substitute dyed dark chocolate brown.
HACKLE—Two dark furnace cock hackles tied in front of wings, covering wing roots.

Dark Spanish Needle (Dry) (Pritt)
HOOK—14.
BODY—Orange silk.
WINGS—A hackle from the darkest part of a brown owl's wing.
HEAD—Bronze peacock herl.

Dark Spanish Needle (Roger Woolley)
HOOK—15.
BODY—Waxed orange or waxed claret tying silk.
HACKLE—Small dark dun feather from shoulder of a starling's wing.
HEAD—Magpie herl, from tail.

Dark Watchet (Wet) (Edmonds and Lee)
HOOK—14–16.
BODY—Mole fur spun on orange and purple tying silks, these silks to show as alternate ribs when wound on.
HACKLE—Small feather from a jackdaw's throat. This is a very dark silvery grey feather.

Dogsbody (Dry) (Harry Powell)
HOOK—14–16.
BODY—Brown tying silk dubbed with camel-coloured dog's hair. Seal's fur dyed this colour can be used as an alternative.
HACKLES (2)—Grizzle (Plymouth Rock) cock's hackle, with a natural red cock hackle wound in the front.
TAIL WHISKS—Three strands from a cock pheasant's tail.

Dun Spider (Wet) (W. C. Stewart)
HOOK—14–15.
BODY—Waxed yellow tying silk.
HACKLE—Short fibred dun feather from under-part of a starling's wing, tied in at head and wound halfway down body.

Ermine Moth (Dry) (Revd. Edward Powell)
HOOK—10–14.
TAG—A loop of two-ply orange wool tied in flat, protruding a quarter of an inch beyond the bend of the hook and then cut off so as to make a short fork.
BODY—White rabbit fur.
RIB—One strand of black knitting wool or coarse black thread.
HACKLES—Two large grey speckled partridge back feathers.

Eric's Beetle (Eric Horsfall-Turner)
HOOK—8.

BODY—Bronze peacock over a double thickness of yellow wool, the wool left showing for two turns at rump.
HACKLE—Black hen, two turns.

February Red (Dry) (Roger Woolley)
HOOK—13 long shanked.
BODY—A dark brown quill from the stem of a peacock's tail, dyed dark claret, or dark claret mohair.
WINGS—Two dark brown grizzled (cree) cock hackles tied flat over the back.
HACKLE—Dark blue dun cock, grizzled if possible.
Hackled Pattern:
BODY—as for above.
HACKLE—A dark rusty blue cock.

Gold Ribbed Hare's Ear (Wet)
HOOK—14–16.
WHISKS—Three long strands of hare's body fur.
BODY—Dark hare from the root of a hare's ear spun on yellow silk.
RIB—Fine flat gold tinsel.
HACKLE—The longer fibres of the body fur picked out with a dubbing needle, or the longer body hairs spun on the tying silk and worked as a hackle.
Wings (if required) from a starling's wing primary feathers.

Governor (Dry) (T. C. Hofland)
HOOK—14.
TIP—Scarlet silk.
BODY—Bronze peacock herl.
WINGS—From hen pheasant wing primary feathers.
HACKLE—Light red or ginger cock's hackle.

Grannom (Dry) (Revd. E. Powell)
HOOK—14.
TYING SILK—Green.
BODY—Mole fur dyed in picric acid.
HACKLES—Two partridge brown back feathers, as light as possible.

Grannom (Dry) (Cliff Henry)
HOOK—15.
TYING SILK—Yellow.
BODY—Hare's ear ribbed fine gold wire—two or three turns of green silk at tail.
WINGS—Fawn mallard primaries tied sloping back over body.
HACKLE—Light ginger cock, short in the flue.

Gravel Bed (Dry) (Ronalds)

HOOK—12.
BODY—Dark dun or lead coloured silk.
WINGS—From a woodcock's wing feather, tied sloping over body.
HACKLE—A black cock's hackle, fairly long in fibre, wound twice only. The fibres of the hackle should be about twice the length of the hook.

Gravel Bed (Dr. Pryce-Tannatt)

BODY—Fibres from the blue-grey feather of a heron's wing.
HACKLES (2)—A black cock's hackle wound as for the Ronalds pattern, with a partridge brown back feather wound in front.

Great Red Spinner (J. R. Harris)

HOOK—12–13.
TAIL WHISKS—Fibres from a dark rusty dun or red cock hackle.
BODY—Dark red or claret seal's fur on claret tying silk.
RIB—Gold wire.
HACKLE—Dark rusty dun cock, tied half spent.

Great Red Spinner (Roger Woolley)

TAIL WHISKS—Fibres from a red game cock hackle.
BODY—Red seal's fur.
RIB—Gold wire.
WINGS—Tips of two medium blue dun cock hackles tied spent, or medium blue dun cock hackle fibres tied spent.
HACKLE—Red cock hackle.

Green Caterpillar (Dry) (Courtney Williams)

HOOK—12–14.
BODY—Emerald green wool.
HACKLE—Stiff emerald green cock hackle wound length of body and clipped short.

Greenwell's Glory (Dry or Wet) (Canon William Greenwell)

HOOK—14–16.
BODY—Yellow silk waxed with black cobbler's wax to give it a greenish olive hue.
WINGS—From a blackbird's wing feathers.
HACKLE—Light Coch-y-Bonddu or furnace hackle.
  A rib of fine gold wire is usually added nowadays, but the above is the original pattern.
Hackled Pattern:
TAIL WHISKS—Fibres from a furnace cock's hackle.
BODY—Yellow tying silk—well waxed.
HACKLES (2)—A light coch-y-bonddu or furnace cock's hackle and a medium blue dun cock's hackle.

Grey Duster (Dry) (Courtney Williams)
HOOK—12–14, larger for lakes.
TYING SILK. Brown.
BODY—A dubbing of light rabbit fur, sometimes with a little of the blue under-fur added.
HACKLE—A badger cock's hackle with a well-marked black centre and white list.

Hawthorn (Dry) (Courtney Williams)
HOOK—12.
BODY—Black ostrich herl, thin.
WINGS—From a starling's wing feather.
HACKLE—Black cock's hackle, fairly long in fibre.

Hawthorn (Dry) (Roger Woolley)
BODY—Two strands from a black turkey tail feather, tied and wound so that the shiny black quill shows up well. The two ends of the strands are tied back to represent the legs of the fly.
HACKLE—Black cock's hackle.
WINGS—From a jay's primary wing feathers, as pale as possible.

Houghton Ruby (Dry) (William Lunn)
HOOK—16.
TYING SILK—Crimson.
TAIL—Three fibres from a white cock hackle.
BODY—Rhode Island hackle stalk dyed crimson.
WINGS—Two light blue dun hen tips, using feathers taken from the breast or back, set on flat.
HACKLE—Natural bright red.

Imperial (Dry) (Oliver Kite)
HOOK—15 or 14.
TYING SILK—Purple.
BODY—Undyed heron primary herls.
THORAX—Same herls doubled and redoubled.
RIB—Gold wire.
TAILS—Grey brown in spring, later honey dun.
HACKLE—Honey dun.

Iron Blue (Dry)
HOOK—16.
BODY—Stripped peacock quill dyed purple or claret.
WINGS—Cock blackbird, or starling dyed inky blue.
HACKLE—Dark blue dun cock's hackle.
TAIL WHISKS—Fibres from a pale blue dun cock's hackle.
OR

BODY—Dark peacock quill dyed olive.
WINGS—as above.
HACKLE—Dark brown olive cock's hackle.
WHISKS—Fibres from a rusty dun cock's hackle.
Hackled Pattern:
BODY—Purple or claret quill.
HACKLE—Dark blue dun cock's hackle.
TAIL WHISKS—Fibres of pale blue dun cock's hackle.

Dark Watchet (North Country wet pattern)
BODY—Mole fur dubbed on orange and purple tying silk, which should be twisted together to make alternate ribbings.
HACKLE—Small feather from a jackdaw's throat.

John Storey (Dry)
HOOK—Any size.
TAIL—Fibres of red cock's hackle.
BODY—Thin peacock herl.
RIB—Scarlet silk.
HACKLE—Red cock hackle.
WINGS—Grey mallard flank feather fibres.

John Storey (Eric Horsfall-Turner)
HOOK—16.
BODY—Bronze peacock.
HACKLE—Dark red cock.
WING—Forward-tied tuft of mallard breast feather.

July Dun (Dry) (Roger Woolley)
TAIL WHISKS—Medium olive cock hackle fibres.
BODY—Heron herl dyed yellow.
RIB—Gold wire.
WINGS—From a starling's wing primary feathers, as dark as possible.
HACKLE—Medium olive cock's hackle.
Hackled Pattern:
BODY AND TAIL WHISKS—as for winged fly.
HACKLE—Darkish dun cock's hackle.

Large Green Dun (Dry) (John Veniard)
HOOK—12-14.
TAIL WHISKS—Fibres from a dark dun cock's hackle.
BODY—Greenish grey seal's fur or herl from a heron's wing quill dyed light green.
RIB—Dark brown or black tying silk.
HACKLES (2)—Dyed green cock's hackle, with a longer fibred dyed green grizzle cock's hackle in front.

Large Green Dun (Leonard West)

TAIL WHISKS—Fibres from a green cock's hackle.
BODY—Green quill.
RIB—Gold wire.
THORAX—Dark green dubbing or herl.
WINGS—From a starling's wing primary feathers.
HACKLE—Dyed green cock hackle.

Large Green Spinner (John Veniard)

HOOK—13–14.
TAIL WHISKS—Two fibres from a dark dun cock's hackle, fairly long.
BODY—Stripped peacock quill dyed green, well marked.
WINGS—Grizzle cock hackle tips tied spent, not too heavily marked.
HACKLE—Cock's hackle dyed green, short and stiff fibred.

Last Hope (Dry) (John Goddard)

HOOK—17 or 18 up eye fine wire.
TYING SILK—Pale yellow.
BODY—Goose primary or condor herls buff colour.
HACKLE—Cream cock very short in flue.
TAILS—Honey dun cock, six fibres.

Little Brown Sedge (Dry) (Courtney Williams)

HOOK—14.
BODY—Orange tying silk dubbed with light brown wool.
RIB—Fine gold wire.
BODY HACKLE—Short fibred natural red cock hackle from shoulder to tail.
WINGS—From wing feathers of Rhode Island Red hen.
FRONT HACKLE—Natural red Cock's hackle, tied in front of wings over wing roots.

Little Marryat (Dry) (Skues)

HOOK—15–16.
TYING SILK—White or pale straw.
TAIL WHISKS—Creamy dun.
BODY—Cream fur from a baby seal.
WINGS—Pale starling wing feather.
HACKLE—Creamy dun.

Little Red Sedge (Dry) (Skues)

HOOK—14.
TYING SILK—Hot orange, well waxed with brown wax.
BODY—Dark fur from hare's ear.

RIB—Fine gold wire, binding down the body hackle.
BODY HACKLE—Short fibred, deep red cock's hackle from shoulder to tail.
WINGS—Brown hen's wing quill rolled and tied sloping well back over body.
FRONT HACKLE—Deep red cock tied in front over wing roots. Five or six turns.

Lunn's Yellow Boy (Spinner) (William Lunn)
HOOK—13-16.
TYING SILK—Light orange.
TAIL WHISKS—Pale buff cock hackle fibres.
BODY—White hackle stalk dyed medium yellow.
WINGS—Two light buff cock hackle tips put on flat.
HACKLE—Light buff.

Lunn's Particular (Spinner) (William Lunn)
HOOK—14-16.
TYING SILK—Crimson.
TAIL WHISKS—Fibres from a large Rhode Island Red cock hackle.
BODY—Undyed hackle stalk of a Rhode Island Red.
WINGS—Two medium blue dun cock hackle tips put on flat.
HACKLE—Medium Rhode Island Red cock hackle.

March Brown (Wet) (John Veniard)
Winged Wet Fly (Northern dressing):
HOOK—13-11.
TAIL—Fibres from a brown speckled partridge hackle or tail feather.
BODY—Sandy fur from a hare's neck, or brown seal's fur.
RIB—Yellow silk or gold wire.
HACKLE—Brown partridge.
WINGS—Fibres from a speckled partridge tail feather.
Winged Wet Fly—Male:
TAIL—Two strands of brown speckled partridge feather.
BODY—Dark hare's fur from ear.
RIB—Yellow tying silk or gold wire.
HACKLE—Brown speckled partridge, or brown grizzled (cree) hen's hackle.
WINGS—Mottled secondary feather from hen pheasant's wing.
Winged Wet Fly—Female:
TAIL and HACKLE as male.
BODY—Ginger hare's fur from neck.
RIB—Gold wire.
WINGS—Same as for male but lighter in colour.
M

Hackled Wet Fly:
   TYING SILK—Hot orange.
   BODY—From a hare's poll, dyed hot orange.
   HACKLE—A snipe's rump feather.

March Brown Spider
   TAIL—Two strands from a speckled partridge tail.
   BODY—Dark hare's ear, mixed with a little claret wool or seal's fur.
   RIB—Yellow or primrose tying silk.
   HACKLE—Brown speckled partridge, fairly long in fibre.

March Brown Spider (Welsh pattern)
   BODY—As above.
   RIB—Silver.
   HACKLE—As above.
   I understand the name of this pattern to be Petrisen Corff Blewyn Ysgyfarnog.

March Brown Nymph
   TAIL—Two short strands of a cock pheasant's tail, or fibres of brown mallard shoulder feather.
   BODY—Herls from a cock pheasant's tail.
   RIB—Gold wire.
   THORAX—Hare's ear fur at shoulder.
   WING CASES—From a woodcock wing feather (sometimes omitted).
   LEGS—One turn of a small brown speckled partridge hackle.
Winged Dry Fly (Male):
   TAIL—Fibres from a brown grizzled (cree) cock's hackle.
   BODY—Hare's ear fur, mixed with a very small amount of yellow seal's fur.
   RIB—Yellow tying silk.
   WINGS—Darkish mottled hen pheasant or cock pheasant wing quill, tied upright.
   HACKLE—Brown grizzled (cree) cock's hackle—stiff and bright.
Winged Dry Fly (Female):
   Tied as for the male, but with wings of a lighter shade.
Hackled Dry Fly:
   TAIL—Stiff fibres from a brown grizzled (cree) cock's hackle.
   BODY—As for winged dry flies.
   HACKLES—Same as for winged dry flies, but with the addition of a bright red game cock hackle (short in fibre) tied behind the cree hackle. Alternatively, brown speckled partridge hackle wound mixed with a darkish bright blue dun cock's hackle.

March Brown variations
Silver March Brown:
- TAIL—Two fibres from a brown speckled partridge hackle or speckled partridge tail feather.
- BODY—Flat silver tinsel.
- RIB—Oval silver.
- WINGS—From a hen pheasant's mottled secondary wing feather.

Purple March Brown:
A pattern of recent innovation, very popular in Scotland and practically always fished wet. The only deviation from the standard pattern is its purple wool or seal's fur body, ribbed with yellow silk or gold wire.

Mayflies (John Veniard)
Spent Mayfly:
- HOOK—10–12 long shank.
- TAIL—Three cock pheasant tail fibres.
- BODY—White floss silk or natural raffia.
- WINGS—Dark blue dun hackle points tied "spent".
- HACKLE—Badger cock hackle.

Fan Wing Mayfly:
- HOOK—10–11, long shank.
- TAIL—Three fibres from a cock pheasant tail.
- BODY—Yellow raffia.
- RIB—Oval gold tinsel.
- WINGS—Duck breast feathers back to back, dyed yellow.
- HACKLES—1st, yellow cock hackle.
  2nd, grey partridge back feather, or a grizzle cock hackle.

Shaving Brush Mayfly:
This fly derives its name from the peculiar wing formation, and is a good imitation of the hatching fly. The "Barret's Brown" is also of the same type, and the "Grey Wulf", both dressings being given here.
- HOOK—10–11, long shank.
- TAIL—Three long black cock hackle fibres.
- BODY—Kapok.
- RIB—Black tying silk.
- WINGS—A large grizzle hackle wound and divided to slope forward.
- HACKLE—Black cock's hackle.

Barret's Brown Mayfly:
- HOOK—10–11, long shank.
- TAIL—Three long black cock hackle fibres.
- BODY—Brown raffia.
- RIB—Oval gold tinsel.
- WINGS—Fibres from a brown bucktail, divided and sloping forward.
- HACKLE—No. 1, medium olive cock's hackle.
  No. 2 (nearest to eye), hen pheasant neck feather.

Grey Wulf Mayfly (Hair Winged):
:   HOOK—10–11, long shanked.
    TAIL—Fibres from a brown bucktail.
    BODY—Blue-grey seal's fur.
    WINGS—Brown fibres from a bucktail, divided and sloping forward.
    HACKLE—Blue dun cock's hackle.

    Tied on a size 15 hook, and using fibres from a brown barred squirrel tail for the wings, this makes an excellent representation of smaller hatching nymphs.

French Partridge Mayfly:
:   This is a mayfly of the "Straddle Bug" type, as also is the pattern immediately following.
    HOOK—10–12, long shanked.
    TAIL—Three fibres from a cock pheasant tail.
    BODY—Natural raffia ribbed with oval gold tinsel.
    BODY HACKLE—Olive cock's hackle.
    FRONT HACKLE—French partridge breast feather.

Straddle-Bug Mayfly (Summer Duck):
:   HOOK—10–12, long shank.
    TIP—Very fine gold oval or twist tinsel.
    TAIL—Two or three fibres from a black cock's hackle.
    BODY—Natural raffia.
    RIB—Brown tying silk.
    HACKLES—No. 1. Orange cock's hackle.
    No. 2. Brown speckled summer duck feather.
    HEAD—Bronze peacock herl.

Spent Mayfly:
:   This is a personal favourite which I have used with much success during the height of the hatch.
    HOOK—10–11, long shanked and of fine wire.
    TAIL—Three fibres from a cock pheasant tail.
    BODY—Natural raffia.
    RIB—Oval silver tinsel.
    WINGS—A very stiff black or dark blue dun cock's hackle wound in the normal way, the fibres then being split into two equal halves and tied "spent".
    HACKLE—None. I rely on the stiffness of the wings to float the fly, but if a hackle is preferred I would recommend a short fibred stiff badger hackle.

Mayflies (Wet and Nymphs) (John Veniard)

Mayfly Nymphs:
:   These should be tied fairly heavily as the mayfly nymph is a good-sized insect. There are many patterns but the following three should meet most requirements.

No. 1. HOOK—11-9, long shanked.
  TAIL—Three short strands from a cock pheasant's tail feather.
  BODY AND THORAX—Brown olive seal's fur.
  RIB—Oval gold tinsel.
  WING CASE—From a hen pheasant's tail feather. Dark as possible.
  LEGS—Brown partridge back feather.
No. 2. TAIL—As for No. 1.
  BODY—Three turns at tail end of dirty yellow seal's fur. The remainder and thorax of brown olive seal's fur.
  RIB—Yellow silk, thicker than the normal tying silk.
  LEGS—A mottled feather from a hen pheasant's neck.
No. 3. TAIL—As for Nos. 1 and 2.
  BODY AND THORAX—Pale buff seal's fur.
  RIB—Oval gold tinsel.
  WING CASE—As for Nos. 1 and 2.
  LEGS—Dark brown grouse hackle.

Wet Mayflies:
More often than not these are used at the commencement of the rise while the fly is hatching, and should therefore be fished semi-submerged.

No. 1. HOOK—11-9, long shanked.
  TAILS—Three strands from a cock pheasant's tail.
  BODY—Dyed yellow lamb's wool.
  RIB—Oval gold tinsel.
  LEGS—Hen pheasant neck feather dyed yellow, or undyed.
  WINGS—Speckled grey mallard feather dyed yellow.
No. 2. TAILS—As for No. 1.
  BODY—As for No. 1.
  RIB—As for No. 1.
  LEGS—Ginger cock's hackle, dyed yellow.
  WINGS—Speckled grey mallard feather dyed greenish yellow.
No. 3. TAILS—As for Nos. 1 and 2.
  BODY—Buff yellow floss silk.
  RIB—Oval gold tinsel.
  LEGS—Brown grouse hackle.
  WINGS—Grizzle cock hackle dyed greenish yellow.

Green Drake Upright (Dry) (David Jacques):
  HOOK—A "Mayfly" hook, iron as light as possible.
  TYING SILK—Olive or straw colour.
  BODY—Build up shape with floss silk, then cover with natural raffia.
  HEAD HACKLE—Two short stiff cock hackles dyed green drake.
  TAIL HACKLE—Two or more long cock hackles, as stiff as possible, same shade as head hackles.
Grey Drake Upright (David Jacques):
  As above, but the hackles to be dyed medium grey.

Nevamis Mayfly (Dry) (John Goddard):
   HOOK—Long shank fine wire No. 8.
   TYING SILK—Yellow.
   BODY—Baby seal fur wound thickly—body hackle large honey cock tied in at tail and wound to shoulder and then clipped to $\frac{1}{4}$ inch of body at shoulder, sloping to $\frac{1}{8}$ inch at tail.
   RIB—Oval gold tinsel.
   WINGS—V-shape hackle fibre wings using a pale blue dun cock (1 inch fibres)
   HACKLE—Small furnace cock ($\frac{1}{2}$ inch fibres).
   TAILS—Three long pheasant tail fibres.

Medium Sedge (Dry) (Halford)
   HOOK—11-13.
   BODY—Unstripped condor herl dyed medium cinnamon. (Or cinnamon turkey tail fibres as an alternative.—Veniard.)
   BODY HACKLE—Short fibred ginger cock hackle.
   WINGS—Well-coloured red hen wing quill.
   FRONT HACKLE—Ginger cock wound in front of wings over wing roots.

Mole-fly (Dry)
   Although predominantly popular in Europe, particularly in France, this fly is of English origin, taking its name from the river Mole in Surrey. Very effective on slow-flowing rivers, where its forward style of winging enables it to cock up very nicely. It will take fish during a hatch of Olives, Mayflies or Sedges.
   HOOK—14-16.
   BODY—Dark olive tying silk.
   RIB—Gold wire.
   WINGS—Mottled hen pheasant tied forward.
   HACKLE—Red/black (furnace) wound from shoulder to tail.

Oak-fly (Dry) (Roger Woolley)
   HOOK—11.
   BODY—Orange raffia.
   RIB—Dark quill from stem of a peacock tail, stripped.
   WINGS—Woodcock wing quill, or two tips of two small dark grizzled cock hackles, tied low over back.
   HACKLE—Furnace cock.

Olive Dun (Dry) (Cliff Henry)
   HOOK—16 up eye.
   TYING SILK—Green.
   TAIL—Grey-blue whisks.
   BODY—Hen hackle stalk from light olive cape.

WINGS—Mallard duck quill feather.
HACKLE—Pale gingery olive.

Olive Quill (Dry)

HOOK—14–16.
TAIL WHISKS—Fibres from a medium olive cock hackle.
BODY—Stripped peacock quill from "eye" part of tail, dyed olive.
WINGS—Dark starling wing quill feather.
HACKLE—Medium olive cock hackle.

Olive Upright (Wet)

HOOK—14–15.
TAIL WHISKS—Fibres from an olive cock hackle.
BODY—Stripped peacock quill from "eye" of tail, dyed yellow.
HACKLE—Softish olive cock, or olive hen hackle.

Orange Quill (Dry) (Skues)

HOOK—13–14.
TYING SILK—Orange.
TAIL WHISKS—Fibres from a natural bright red cock hackle.
BODY—Pale condor quill, stripped and dyed hot orange. (Condor is difficult to strip, and I have found stripped ostrich herl equally effective.—Veniard.)
WINGS—Pale starling wing quill, rather full.
HACKLE—Natural bright red cock hackle.

Pale Evening Dun (Dry) (Oliver Kite)

HOOK—15.
TYING SILK—White.
BODY—Grey goose, primary herls.
THORAX—Same herls doubled and redoubled.
TAILS—Cream cock.

Partridge and Orange (Dry)

HOOK—14–15.
BODY—Orange silk.
HACKLE—Speckled feather from a partridge's back.
RIB (optional)—Gold wire.

Pheasant Tail (Nymph) (Frank Sawyer)

HOOK—16–12 D/E.
Material for tails, body and wing cases—Cock pheasant centre tail fibres, and fine copper wire. The cock pheasant tail should be from a mature bird with the edges of the tail as deep a red as possible. The wire should also approximate this colouring.

Construction: The copper wire is wound on to the hook to form a core for the body, and should represent the shape of the nymph, i.e. a hump to represent thorax, and then tapering to the tail. This acts as ballast and general appearance.

Take four fibres from the cock pheasant tail, and tie them in with wire so that the tips of the fibres form the tail whisks. Now wind both wire and feather fibres evenly up to the eye of the hook. Fasten with one turn of the wire, then lap the feather fibres backwards—forwards—then backwards again, to form the characteristic nymph "hump" at the thorax. The wire is not used during the lapping, but should be kept separate from the fibres and just brought backwards or forwards as the case may be, to secure each lap of the fibres. The finish should be made with two half hitches of the wire behind the eye of the hook.

No hackle is used, and it will be observed that it is possible to make this pattern without the use of tying silk either! In spite of its simplicity this is one of the most popular and successful patterns ever evolved.

Pheasant Tail (Spinner) (Skues)

HOOK—14–16.
TYING SILK—Hot orange.
TAIL WHISKS—Two or three strands of honey dun cock spade feather.
BODY—Two or three strands of rich-coloured ruddy fibres from centre feather of a cock pheasant's tail.
HACKLE—Rusty or sandy dun cock, bright and sharp.

Port Spinner (Cliff Henry)

HOOK—14–16.
TYING SILK—Crimson.
TAIL—Pale blue cock fibres.
BODY—Hen hackle stalk from dark Rhode Island Red. (Port wine colour.)
WINGS—Pale blue dun hackle points tied "spent".
HACKLE—Two turns of light olive.

Poult Bloa (Wet) (Traditional)

BODY—Yellow silk dubbed with fur from a red squirrel.
HACKLE—Slaty blue feather (poult) from under the wing of a young grouse.

Poult Bloa (Roger Woolley)

Same dressing as above, but using ginger hare's fur, or omitting the fur altogether.

Purple Dun (Dry) (J. R. Harris)
HOOK—14–15.
TAIL—Fibres from a dark blue dun cock's hackle.
BODY—Heron herl dyed purple, or mole's fur spun on purple tying silk.
HACKLES (2)—Black cock's hackle with a dark blue dun wound through it.

P.V.C. Nymph (John Goddard)
HOOK—16–12 D/E.
TYING SILK—Yellow.
BODY—The shank of hook is covered with fine copper wire. Start at eye and work along, building up a thorax just behind eye by winding wire more thickly at this point. Continue on to bend of hook and use wire to tie in body materials, then cut.

Begin with a thin strip of clear P.V.C. followed with three golden olive condor herls. The points of these should protrude $\frac{1}{8}$ inch at rear to form tails. These are then wound over wire to form body up to eye. Tie in with yellow silk at eye and wind silk down to back of thorax, then wind in P.V.C. from tail over body up to back of the thorax and tie in and trim. Take tying silk back to eye and tie in three blackish pheasant tail fibres; these are then doubled and redoubled over top of thorax to form hump to imitate wing cases.

To imitate the translucence of some nymphs, sparsely wind silver tinsel over the herl of the body but under the P.V.C.

Red Ant (Dry) (Wm. Lunn)
HOOK—15–17.
TYING SILK—Deep orange.
BODY—Wind tying silk to form shape of ant's body.
WINGS—Fibres from a large white cock's hackle tied in to lie close over body. Cut off tips to give required length.
HACKLE—Short fibred stiff bright red game cock's hackle.

Red Quill (Dry)
HOOK—14.
TAIL WHISKS—Three fibres from a natural bright red cock's hackle.
BODY—Stripped quill from "eye" part of peacock tail, ensuring that the well-marked light and dark bands prevail.
WINGS—From starling's wing quill.
HACKLE—Natural bright red cock's hackle. Alternative: Pale blue dun cock's hackle, and no wings.

Red Spinner (Courtney Williams)
HOOK—12–14.
TAIL WHISKS—Fibres from a natural bright red cock's hackle.

BODY—Red silk.
RIB—Gold wire.
WINGS—From starling's wing quill.
HACKLE—Natural bright red cock's hackle.

Red Spinner (Roger Woolley)
TAIL WHISKS—Fibres from a bright red cock's hackle.
BODY—(1) Red dyed horsehair on the bare hook.
  (2) Celluloid over red tying silk.
  (3) Red seal's fur ribbed with gold wire.
WINGS—Glassy pale blue dun cock hackle fibres or two small pale blue cock hackle tips, both tied "spent".

Red Spinner (Skues)
(Hackled pattern)
TYING SILK—Hot orange.
TAIL WHISKS—Three fibres from a steely blue dun cock hackle.
BODY—Crimson seal's fur.
RIB—Fine gold wire.
HACKLE—Bright steely blue dun cock's hackle.

Red Spider (Wet) (Stewart)
HOOK—14–15.
BODY—Waxed yellow tying silk.
HACKLE (run half-way down body)—Small feather from outside a landrail's wing, or a small natural red hen's hackle.

Rough Olive (Dry) (Roger Woolley)
HOOK—13–14.
TAIL WHISKS—Dark olive cock hackle fibres.
BODY—Heron's herl dyed olive.
RIB—Fine gold wire.
WINGS—Dark starling or hen blackbird wing quill.
HACKLE—Dark olive cock's hackle.

Sand-fly (Dry) (Ronalds)
HOOK—14.
BODY—Sandy-coloured fur from a hare's neck spun on silk of the same colour or near.
WINGS—From a landrail's wing quill (or substitute) tied rather full.
HACKLE—Ginger cock's hackle. (Hen's hackle for wet fly.)

Sepia Dun (Dry) (Oliver Kite)

HOOK—13.
TYING SILK—Brown.
BODY—Dark undyed heron primary herls.
THORAX—Same herls doubled and redoubled.
RIB—Gold wire.
TAILS—Black cock.
HACKLE—Reddish black cock.

Sherry Spinner (Spinner) (Skues)

HOOK—14.
TAIL WHISKS—Three fibres from a pale honey dun cock's hackle.
BODY—Amber seal's fur. (Skues advocates a mixture of orange, light orange, and green seal's fur, with a small fine amount of fur from a hare's poll added.)
RIB—Fine gold wire.
HACKLE—Palest honey dun, with a darkish centre if possible.

Sherry Spinner (Roger Woolley)

TAIL WHISKS—Pale ginger cock hackle fibres.
BODY—Pale gold quill or gold-coloured floss.
WINGS—Glassy blue dun cock hackle fibres, or two small pale blue cock hackle tips, both tied "spent".
HACKLE—Pale ginger cock's hackle.

Shrimp (Wet) (Eric Horsfall-Turner)

HOOK—11 D/E.
BODY—Orange DRF.
RIB—Silver wire.
HACKLE—Ginger cock palmered evenly along the body.
BACK—Light brown turkey laid from tail to head.

Silver Sedge (Dry) (Courtney Williams)

HOOK—10-14.
BODY—White floss silk.
RIB—Fine silver wire, and a sandy ginger cock's hackle.
WINGS—From a landrail's wing (or substitute) tied low over body.
HACKLE—Sandy ginger cock wound over wing roots.

Snipe Bloa (Wet) (Pritt)

HOOK—14-16.
BODY—Straw-coloured silk.
HACKLE—Small feather from outside of a jack snipe's wing.

Snipe and Purple (Wet)
HOOK—14–16.
BODY—Purple floss or tying silk.
HACKLE—Small feather from outside a jack snipe's wing.

Soldier Beetle (Dry) (Skues)
HOOK—14 D/E.
TYING SILK—Hot orange.
BODY—Bright red-orange seal's fur.
WING CASES—Fibres from a cock pheasant's breast tied in at tail, brought down over body and tied in at head.
HACKLE—Natural red cock, rather sparse.

Stonefly (Roger Woolley)
HOOK—No. 11 long shank.
BODY—(1) A mixture of two-thirds dark olive seal's fur, and one-third dirty yellow seal's fur.
    Rib—Yellow sewing silk.
(2) Dark olive-brown raffia over a foundation of yellow wool, also ribbed with yellow sewing silk.
WINGS—The tips of four dark grizzle cock hackles tied low and flat over the back of the fly, and projecting well beyond the hook.
HACKLE—Grizzle cock dyed brown-olive.

Stonefly (John Veniard)
BODY—A mixture of dark hare's fur/dark yellow seal's fur.
RIB—Yellow silk, close wound at tail end.
WINGS—Four dark dun cock hackles tied low over back and projecting beyond hook.
HACKLE—Dark grizzle cock wound in front of wings.
Alternative wings: Fibres from the dark part of a hen pheasant's wing quill, rolled and tied low over body.

Tup's Indispensable (Dry) (R. S. Austin)
HOOK—13–16.
TAIL WHISKS—Brassy or honey dun cock hackle fibres.
BODY—Yellow silk at tail end, creamy pink dubbing for the remainder.
HACKLE—Brassy or honey dun cock's hackle.

Tup's Indispensable (Wet)
For the nymph imitation, the dubbing should be taken nearer to the tail, and the hackle should be a short fibred blue dun hen's hackle. Only two or three turns of yellow silk should show at tail end.

Water-Hen Bloa (Wet) (Pritt)
HOOK—14.
BODY—Yellow silk, dubbed with the fur of a water-rat (alternatively mole's fur).
WINGS—Hackled feather from the inside of a moor-hen's wing.

Winter Brown (Dry) (Roger Woolley)
HOOK—13 long shanked.
BODY—Dark brown quill from the stem of a peacock's tail, dyed dark brown.
WINGS—Two dark cree (brown grizzle) cock hackles dyed dun, and tied flat along back.
HACKLE—Dark brown dun cock's hackle.
Hackled Pattern:
BODY—As for the winged pattern.
HACKLE—A dark, well-rusted blue dun cock hackle.

Yellow Evening Dun (Dry) (J. R. Harris)
HOOK—14.
TAIL WHISKS—Fibres from a ginger cock's hackle.
BODY—Bright orange floss, or light orange quill.
RIB—Gold wire.
WING—Bunch of fibres from a pale yellow cock's hackle tied so as to incline slightly forward.
HACKLE—Ginger cock's hackle.

Yellow Halo (Dry) (John Goddard)
HOOK—17 or 18.
TYING SILK—Yellow.
BODY—A strip of thin, very pale yellow plastic P.V.C. wound over cream condor herl.
HACKLES—A bright yellow cock hackle is trimmed to within $\frac{1}{8}$ inch of the stalk. Wind in two turns as a hackle. The main hackle of cream cock is then wound over this.
TAILS—Honey dun cock four to six fibres.

Yellow Evening Spinner (J. R. Harris)
HOOK—14.
TAIL—Fibres from a ginger cock's hackle.
BODY—Light orange seal's fur spun on orange tying silk.
RIB—Gold wire.
WING—Honey dun or pale ginger cock hackle fibres tied "spent" or "half spent".
No hackle.

Yellow Sally (Dry) (Courtney Williams)

    HOOK—14.
    BODY—Pale yellow wool, teased out before spinning on to primrose tying silk.
    HACKLE—White cock's hackle dyed pale yellow.

Alternative:
    BODY—Light canary quill or wool.
    HACKLE—Two pale ginger cock's hackles.

Yellow Upright. (Dry) (H.P.B.)

    HOOK—14.
    TAIL WHISKS—Eight to ten fibres from a pale honey blue dun cock's hackle.
    BODY—Belly fur from a half-grown hedgepig, blended with a little rusty yellow seal's fur.
    RIB—Fine gold thread.
    HACKLE—Good quality honey (or honey dun, if available) cock's hackle, plenty of turns to ensure maximum buoyancy.

APPENDIX B
SPECIAL PATTERNS.

BY THE AUTHOR

Like many amateur fly tiers, I have over the years perfected and tied many special patterns for my own personal use. The majority of these have been designed to represent the natural fly of interest at the time as closely as possible, and have been the result of a close study of the natural combined with coloured photographs which are of course essential during winter fly tying sessions when the naturals are unobtainable.

In the appendix of artificial dressings listed in this book will be found several of these patterns which have consistently proved their value over the years to myself and also many of my fly fishing friends. Out of these patterns there are three that warrant special mention as they have proved to be exceptionally killing patterns. These are as follows.

The P.V.C. Nymph

This is a weighted nymph pattern and although the basic material is copper wire to add weight as in Sawyers Pheasant tail nymph, here the resemblance ceases. In my pattern olive coloured condor herl is used to cover the wire and on top of this material the body is again covered with a closely bound strip of olive dyed P.V.C. This relatively new material when tied over the herl gives the body a wonderfully translucent appearance while closely simulates the natural nymph. For added effect I sometimes tie in a sparce rib of narrow silver lurex over the herl but under the P.V.C. which I think increases the translucence, and probably gives the appearance of air bubbles trapped under the skin, which can often be observed in the natural nymph immediately prior to emergence.

This pattern is meant to represent the many different species of Olive nymphs which frequent a lot of our rivers, it can be tied on hooks from size 12 up to 16. These nymphs can be tied to fish at different levels according to the amount of weight provided in the form of copper wire.

The Hatching Olive (Plate 11).

In appearance this is very similar to the previous pattern except the basic copper wire is omitted and it has a sparce hackle added so that it floats in the film tail down in a similar manner to the natural nymph immediately prior to emergence. This pattern is tied to represent the natural on the point of emergence and as such it should always be fished in the surface film, and is therefore basically fished as a dry fly. At certain times trout feed exclusively on these hatching nymphs and will completely ignore the freshly hatched duns. When trout are so engaged the rise form is practically identical to the rise to the dun, and it is only with considerable experience that a slight difference can sometimes be detected. It is at this time that this pattern is indispensable as even an unweighted nymph pattern is usually unacceptable as it will fish an inch or so under the film which is usually too deep to be accepted by the trout.

N.B. Later in the season I often use peacock herl to form the Thorax to give a slightly darker silhouette.

The Last Hope (Plate 12).

This tiny fly has accounted for a lot of big trout. The materials used in the dressing of this pattern are not particularly original, but the secret of its success lies in the method of dressing. It should always be tied on a very small hook size 17 or 18 and it is essential to use a small sharp hackle that is extremely short in the flew. This type of hackle is extremely difficult to obtain so as an alternative a clipped hackle may be used. Finally to aid floatability at least six short tail whisks should be tied in. For the body I have found two or three herls in a dark grey stone colour or alternatively light buff to be ideal, preferably from the breast feather of a Norwegian goose. Condor herl may be used as an alternative.

This pattern was originated to represent the Pale Watery dun, but in practice it has proved extremely effective when any small flies are on the water. I use the lighter bodied buff pattern early in the season and the darker grey pattern from Mid June onwards.

The Polystickle.
By Dick Walker.
An excellent imitation of a sticklebeck.

The Hatching Olive.
By the Author.
To represent the Pond or Lake Olive hatching in the surface film.

The Adult Midge.
By the Author.
May be dressed in various colours and sizes.

Plate 11. Special Patterns.

RED OR GREEN LARVAE.
By the Author.
A killing pattern fished in the prescribed manner

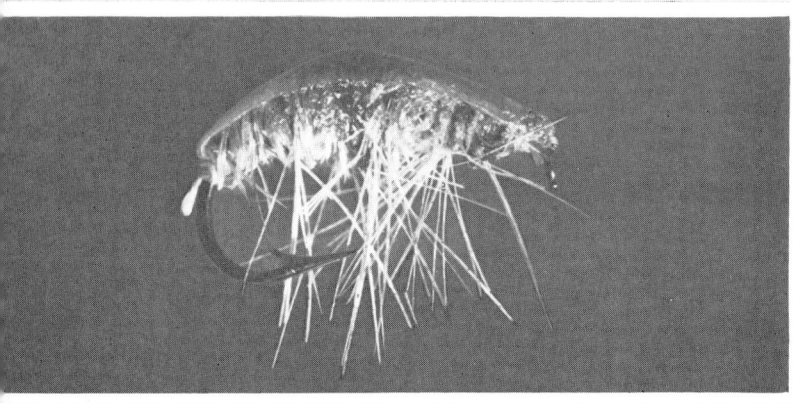

THE SHRIMPER.
By the Author.
A very lifelike pattern to simulate the freshwater shrimp.

THE FLEXIFLY.
By the Author.
A new approach to a general pattern.

THE LAST HOPE.
By the Author.
A tiny artificial—but a big fish killer.

PLATE 12. Special Patterns.

THE G. AND H. SEDGE.
(*side view*).
Developed by the Author and Cliff Henry this has proved itself time and again when any green bodied sedges are on the water.

THE G. AND H. SEDGE.
(*top view*).

OLIVE DUN.
By Cliff Henry.
A superb dry fly when Pond or Lake Olives are on the water.

PLATE 13. Special Patterns.

APPENDIX C

BODY PARTS OF FLIES

ENTOMOLOGICAL NAME	ANGLER'S NAME
Abdomen	The main body below head and thorax
Antennae	Feelers
Anterior leg or wing	Foreleg or forewing
Cerci	Tails
Copulation	The mating act
Costa	Leading edge of wing
Costal process or projection	Spur on leading edge of wing
Cuticle	Skin encasing dun or nymph
Dorsum	Top of main body segments
Femur, Femora	Top part of legs
Forceps	Claspers of male fly
Intercalary veins	Small veins between main veins
Median leg	Middle leg
Oculi	Eyes (compound)
Posterior leg or wing	Back leg or rear wing
Pterostigmatic area	Area near edge of wing apex
Tarsus, Tarsi	Feet or claws
Thorax	Shoulders, chest or wing cases
Tibia	Middle part of legs
Venation	Veining of the wings
Venter	Under part of the main body

BODY PARTS OF FLIES

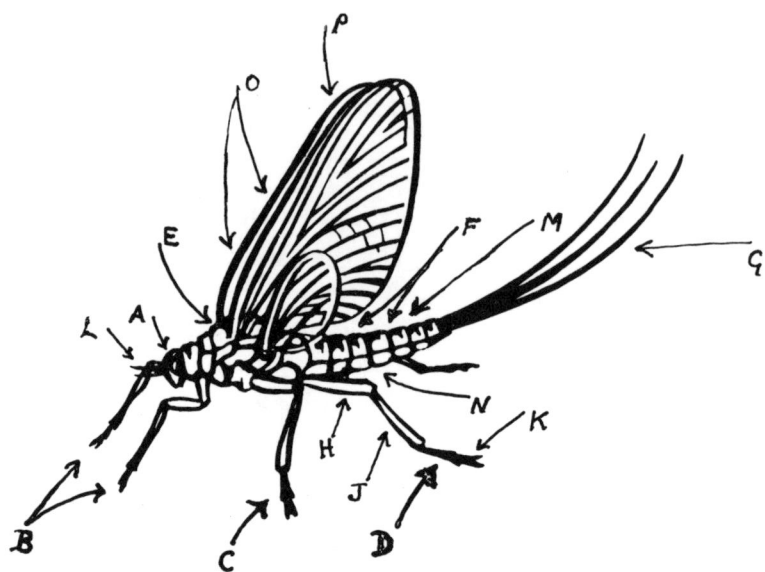

FIG. 51. Entomological names for various parts of an upwinged fly.

A. Oculi.
B. Anterior legs.
C. Median „
D. Posterior „
E. Thorax.
F. Tergites.
G. Cerci.
H. Femur.
J. Tibia.
K. Tarsus.
L. Antennae.
M. Dorsum.
N. Venter.
O. Costa.
P. Pterostigmatic area.

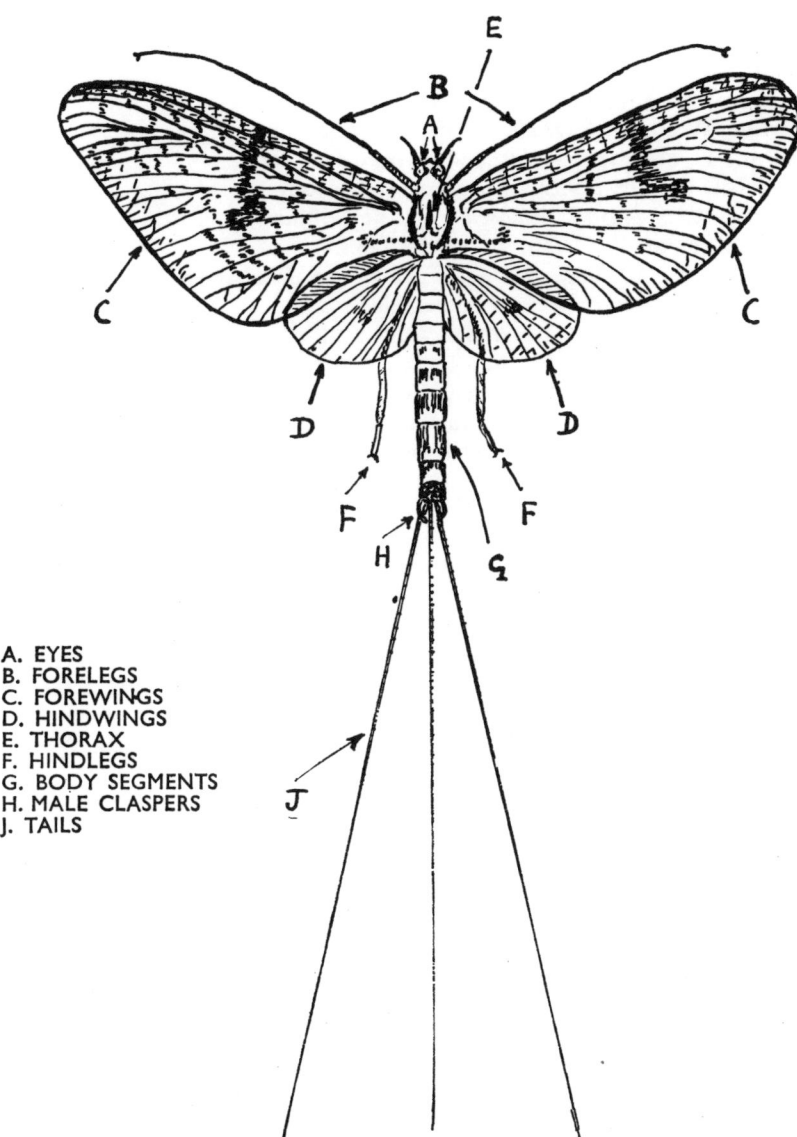

FIG. 52. A Mayfly—showing fisherman's names for various parts.

A. EYES
B. FORELEGS
C. FOREWINGS
D. HINDWINGS
E. THORAX
F. HINDLEGS
G. BODY SEGMENTS
H. MALE CLASPERS
J. TAILS

REFERENCES

Colyer and Hammond. *Flies of the British Isles.* 1951 Warne.
Courtney Williams, A. *Dictionary of Trout Flies.* 1949 Black.
Gledhill, T. *Traps in Two Streams During 1958.* 1960.
Halford, F. M. *Dry Fly Entomology.* 1897 Vinton.
Harris, J. R. *An Angler's Entomology.* 1952 Collins.
Hickin, Norman E. *Caddis.* 1952 Methuen.
Hynes, H. B. N. D.Sc. *Key to the Adults and Nymphs of British Stoneflies.* 1958 Freshwater Biological Assoc. No. 17.
Imms, A. D. *Insect Natural History.* 1947 Collins.
Kimmins, D. E. *Key to the Revised Adults of the British Species of Ephemeroptera.* 1954 Freshwater Biological Assoc. No. 15.
Kite, Oliver. *Nymph Fishing in Practice.* 1963 Jenkins.
Lewis, T. and Taylor, L. R. *Diurnal Periodicity of Flights by Insects.* Royal Entomological Society. Vol. 116, pt. 15, pp. 393–479.
Macan, T. T. M.A., Ph.D. *Key to the Nymphs of the British Species of Ephemeroptera.* 1961 Freshwater Biological Assoc. No. 20.
Macan, T. T. and Worthington, E. C. *Life in Lakes and Rivers.* 1951 Collins.
Mellanby, Helen. *Animal Life in Fresh Water.* 1938 Methuen.
Mosely, Martin E. *British Caddis Flies.* 1939 Routledge.
Mosely, Martin E. *Dry Fly Fisherman's Entomology.* 1920 Routledge.
Novak, K. and Sehnal, F. *Development Cycle of Some Species of the Genus Limnephilus.* 1963 Czechoslovak Institute of Entomology.
Pleskot, G. *Die Periodizität Einiger Ephemeropteren der Schwechat.* Wasser u. Abwasser. 1958. 1–32.
Ronalds, Alfred. *Fly-Fisher's Entomology.* 1868 Longmans Green.
Walker, C. F. *Lake Flies and Their Imitation.* 1960 Jenkins.

INDEX

Abdomen, 25, 50, 109, 120, 132
Acid water, 82
Agile darters, 134, 135, 136, 137, 138, 141, 142, 146
Agriotypus armatus, 130
Alder, 127
Algae, 124, 135
Alkaline water, 82
Anabolia nervosa, 111, 113
Anatomy, 92
Angler's Curse, 31, 103, 143, 146
Annulated, 115
Antennae, 19, 110, 120
Anterior, 25, 26
Ants, 23, 128
Apricot spinner, 30, 65, 96
Arachnidae, 22, 23, 129
Asellus aquaticus, 131
Asellus meridianus, 131
Athripsodes aterrimus, 111, 118
Athripsodes cinereus, 111, 118
Athripsodes nigronervosus, 111, 115
Atrophied, 120, 123
August dun, 31
Autumn dun, 31, 39, 46, 76, 79, 100
Autumn nymph, 141, 146
Autumn spinner, 31, 42, 47, 76, 80, 78, 101
Autopsies, 102, 107, 109, 131

Baëtidae, 31–34
Baëtis, 135
Baëtis atrebatinus, 30, 32, 58, 80, 142, 146

Baëtis bioculatus, 30, 31, 32, 49, 56, 81, 90, 137, 146
Baëtis buceratus, 30, 32, 81, 87, 136, 146
Baëtis genus, 16, 31, 32, 89
Baëtis niger, 30, 31, 32, 53, 89, 86, 136, 146
Baëtis pumilus, 30, 31, 32, 53, 80, 86, 136, 146
Baëtis rhodani, 30, 32, 52, 80, 85, 136, 146
Baëtis scambus, 30, 32, 55, 81, 88, 90, 132, 137, 146
Baëtis tenax, 30, 32, 87, 136, 146
Baëtis vernus, 28, 30, 32, 54, 81, 87, 136, 146
Bees, 130
Beetles, 22, 24, 128
Bibio, 23, 106
Bibio johannis, 23, 105
Bibio marci, 105
Bibionidae, 23
Black Curse, 107
Black-flies, 107
Black Gnat, 23, 105, 106
Black Sedge, 111, 115
Black Silverhorn, 111, 115, 118
Blue dun, 30
Blue-winged olive dun, 18, 27, 28, 29, 30, 37, 38, 44, 46, 60, 73, 83, 89, 91, 92, 94, 95, 97, 99, 100, 139
Blue-winged olive nymph, 135, 137, 146
Blue-winged olive spinner, 60
Blue-winged pale watery dun, 30, 93

Bottom Burrowers, 134, 146
Brachycentrus subnubilus, 111, 117
Brachycercus spp. 31
Brachyptera risi, 124
Broadwings, 31, 134
Broadwing spinner, 31
Brown Sedge, 111, 113
Brown Silverhorn, 111, 118

Caddis-flies, 19, 109
Caddis larvae, 24, 131, 151
Caënidae, 17
Caënis, 31, 39, 47, 103, 134, 143, 146
Caënis spinner, 31, 42, 49
Cantharis livida, 129
Cantharis rustica, 129
Caperer, 111, 112, 113
Carnivorous, 121, 127, 143
Caterpillars, 24
Centroptilum, 135
Centroptilum luteolum, 30, 32, 44, 61, 81, 90, 138, 147
Centroptilum pennulatum, 30, 32, 62, 81, 90, 138, 147
Chironomids, 108
Chironomid larvae, 108
Chironomid pupae, 108
Chloroperla torrentium, 121, 125, 145, 147
Cinnamon Sedge, 111, 114, 118
Claret dun, 31, 38, 46, 74, 100
Claret nymph, 140, 146
Claret spinner, 31, 41, 48, 74, 100
Claspers, 25, 26, 29
Class, 23
Classes, 22
Clegg Thomas, 85
Cloëon, 135
Cloëon dipterum, 30, 32, 49, 64, 95, 141, 147
Cloëon simile, 31
Coch-y-Bonddu, 129
Coleoptera, 22, 128
Colour, 27

Compound Eyes, 25, 110
Copulation, 17, 26, 27, 28, 131
Gordiluridae (The Dung-flies), 107
Costal projection, 40, 50, 119
Crane-flies, 18, 105, 106
Crayfish, 24
Creepers, 21, 121
Cross veins, 64, 65
Crustacea, 22, 23, 131

Daddy-Long-Legs, 105, 106
Damselflies, 23
Dark dun, 30
Dark Olive dun, 30, 32, 40, 45, 58
Dark Olive nymph, 142, 146
Dark Olive spinner, 30, 43, 47, 58
Delophus febrilis, 106
Detritus, 133
Diapause, 114
Dinocras cephalotes, 121, 122, 144, 147
Diptera, 15, 18, 22, 23, 105
Distribution, 82
Ditch dun, 31, 38, 73, 99
Ditch dun nymph, 140, 146
Ditch spinner, 31, 41, 73
Diura bicaudata, 121, 123, 144, 147
Diurnal species, 109
Dorsum, 25, 50
Downlooker, 107
Dragonflies, 23
Dry Fly Entomology, 108
Dun, 16, 35
Dung-flies, 18, 105, 107
Dusky Yellowstreak, 30, 39, 46, 70, 98
Dusky Yellowstreak nymph, 139, 146
Dusky Yellowstreak spinner, 30, 42, 48, 70

Early Brown, 121, 124
Early Brown nymph, 145, 147

Ecdyonurus, 135
Ecdyonurus dispar, 31, 76, 80, 100, 141, 146
Ecdyonurus genus, 103
Ecdyonurus insignis, 31, 49, 78, 80, 101, 142, 146
Ecdyonurus torrentis, 31, 49, 77, 101, 141, 146
Ecdyonurus venosus, 31, 80, 81, 102, 142, 146
Ecology, 82
Egg ball, 92
Eggs (Diptera), 18, 19
Eggs (Ephemeroptera), 16
Eggs (Plecoptera), 21, 22
Eggs (Trichoptera), 20
Emergence, 16
Ephemera, 23
Ephemera danica, 30, 51, 83, 84, 142, 146
Ephemera lineata, 83, 84
Ephemera vulgata, 30, 83, 84, 142, 146
Ephemerella, 135
Ephemerella ignita,. 30, 60, 81, 91, 137, 146
Ephemerella notata, 30, 49, 59, 80, 90, 137, 146
Ephemeridae, 23
Ephemeroptera, 15, 22, 23, 24, 35
Eric's Beetle, 129
Evolution, 82
Eyes, 25

False March Brown, 31
Family, 23
February Red, 121, 124
February Red Nymph, 145, 147
Female (Ephemeroptera), 26
Femur, 26, 96, 97
Filaments, 132, 134
Flat nymphs, 133, 135
Flat-winged Flies, 15–18, 105
Flight, 28
Fly Fisher's Entomology, 116, 131

Forceps, 26, 29
Free-swimming, 19
Freshwater Shrimp, 131

Gammarus pulex, 131
Genera, 23
Genitalia, 34, 36
Genus, 23
Ghostwing, 128
Gill plates, 134
Gills, 132, 134
Gnats, 18, 22, 105
Goera pilosa, 111, 116
Golden spinner, 30, 57, 90
Grannom, 111, 117
Grasshoppers, 23, 130
Gravel bed, 107
Great Dark Drone, 130
Great Red Sedge, 111, 112
Great Red Spinner, 31, 42, 47, 77, 78, 80, 97, 101, 102
Green Drake, 30
Green Lacewing, 128
Greentail Fly, 117
Grey Drake, 84
Grey Flag, 111, 115, 117
Grey Sedge, 111, 113
Ground Beetle, 129
Grouse and Green, 111
Grouse Wing, 111, 118

Habrophlebia, 135
Habrophlebia fusca, 31, 73, 81, 99, 140, 146
Halesus digitatus, 111
Halesus radiatus, 111, 112
Halford, F. M., 108, 114, 127
Hard-winged Flies, 15–21
Harpalus ruficornis, 129
Harris, J. R., 95, 117, 139, 142
Hatching nymph, 96
Hatching pupae, 118, 108
Hawthorn, 195
Head, 25
Hemiptera, 22

Heptagenia, 135
Heptagenia lateralis, 30, 49, 70, 98, 139, 146
Heptagenia sulphurea, 30, 49, 68, 81, 98, 139, 146
Hexatoma fuscipennis, 107
Hilara maura, 106
Hindwings, 80
Horse-flies, 195
Horsfall Turner, Eric, 129
Host, 131
House-fly, 18–19, 105
Hydropsyche augustipennis, 114
Hydropsyche contubernalis, 111, 116
Hydropsyche instabilis, 111, 115
Hydropsyche ornatula, 111
Hydropsyche pellucidula, 111, 115
Hymenoptera, 23, 128, 130
Hynes, H. B. N., 120

Ichneumon-fly, 130
Imago, 17
Immature nymph, 133
Induced take, 131
Insecta, 23
Intercalary veins, 33, 40
Iron Blue dun, 29, 30, 31, 32, 37, 40, 46, 53, 74, 86, 93, 99
Iron Blue nymph, 136, 146
Iron Blue spinner, 30, 43, 44, 47, 53, 87
Isoperla grammatica, 121, 123, 144, 147

Jacques, David, 92
Jenny spinner, 44, 54, 87, 90, 99
Joinings, 50
July dun, 30, 88
June Bug, 129

Keys, 35
Kite, Major Oliver, 86, 95, 131, 133

Labial, 110
Laboured swimmers, 134, 135, 138, 140, 146

Lacewings, 23, 127
Large Amber spinner, 30, 43, 47, 63, 94
Large Brook dun, 31, 38, 46, 77, 101, 142
Large Brook nymph, 141, 146
Large Brook spinner, 31, 42, 47, 77, 78, 101
Large Cinnamon Sedge, 20, 111, 112
Large Dark Olive dun, 30, 32, 40, 45, 52, 85, 97
Large Dark Olive nymph, 136, 146
Large Dark Olive spinner, 30, 43, 47, 52, 86, 88
Large Green dun, 31, 39, 45, 78
Large Green nymph, 142, 146
Large Green spinner, 31, 42, 48, 79, 101, 102
Large Red Sedge, see Great Red Sedge
Large Red spinner, 30, 53, 59
Large Spring Olive, 30
Large Spurwing dun, 30, 32, 40, 45, 62, 90, 93
Large Spurwing nymph, 138, 147
Large Spurwing spinner, 44, 63
Large Stonefly, 121, 122,
Large Stonefly nymph, 132, 144, 147
Larvae (Diptera), 18–19, 107
Larvae (Trichoptera), 20
Larvulae, 132
Late March Brown, 31, 38, 46, 80, 77, 101, 192
Late March Brown spinner, 78
Late March Brown nymph, 142, 146
Leaf Beetle, 126
Legs, 25
Leg spurs, 110
Lepidoptera, 22, 129
Lepidostoma hirtum, 111, 117
Leptocerus aterrimus, 111
Leptocerus cinereus, 111

INDEX

Leptocerus nigronervosus, 111
Leptophlebia, 135
Leptophlebia marginata, 31, 49, 75, 81, 100, 140, 146
Leptophlebia vespertina, 31, 49, 74, 100, 140, 146
Leuctra fusca, 121, 125, 145, 147
Leuctra geniculata, 121, 124, 145, 147
Leuctra hippopus, 121, 125, 145, 147
Lichen, 124
Life cycle (Ephemeroptera), 16–17, 103
Limnephilus lunatus, 111, 114
Limnophilus lunatus, 111
Little Amber spinner, 30, 43, 62
Little Claret spinner, 30, 54, 87
Little Sky Blue dun, 30, 92
Little Yellow May dun, 30
Lunn, W. J., 112
Lunn's Yellow Boy, 90, 94

Macan, Dr. T. T., 142
Male claspers, 93
Male (Ephemeroptera), 26
Male spinners, 44, 92, 93
Mandibles, 134
Marbled Sedge, 111, 112, 116
March Brown dun, 30, 38, 46, 66, 77, 80, 83, 96, 101, 102
March Brown nymph, 139, 141, 146
March Brown spinner, 30, 42, 47, 67, 97
Mating, 28, 93, 106, 121
Mature nymph, 16, 133
Maturity, 121
Maxillary, 19, 110, 116
Mayfly, 18, 23, 30, 47, 51, 83, 84, 120, 133
Mayfly nymph, 132, 133, 134, 142, 146
Mayfly spinner, 49, 51
Median, 25
Medium Olive dun, 28, 30, 32, 40, 45, 54, 87, 90, 95

Medium Olive nymph, 136, 146
Medium Olive spinner, 30, 43, 47, 54, 88
Medium Sedge, 11, 116
Medium Stonefly, 121, 123
Medium Stonefly nymph, 144, 147
Megaloptera, 22, 127
Melasoma populi, 126
Metamorphosis, 17
Midges, 18, 22, 24, 105, 107, 108
Minnows, 24
Monkswood, 106
Moonlight dodging, 121
Mosley, M. E., 92
Mosquitoes, 18, 105
Moss Creepers, 134, 135, 137, 146
Moths, 22, 109, 129
Moult, 18
Murragh, 111
Mystacides azurea, 111, 118
Mystacides longicornis, 111, 118
Mystacides nigra, 111, 118

Needle-fly, 21, 121, 125
Needle-fly nymph, 145, 147
Nemoura cinerea, 121, 125, 147
Nemurella picteti, 121, 125, 147
Neuroptera, 23, 127
Nocturnal species, 109
Non-case making, 115
Northern Bustard, 111
Novak and Sehnal, 114
Nymphal case, 16
Nymph fishing in practice, 131, 133
Nymphs (Ephemeroptera), 16, 132, 151
Nymphs (Plecoptera) 21, 132

Oak-fly, 107
Oculi, 25
Ocydromia glabricula, 106
Odonata, 23
Odontocerum albicorne, 111, 113, 117
Olive dun, 31

Olive Upright dun, 30, 39, 45, 67, 97
Olive Upright nymph, 141, 146
Opaque, 35
Orange-fly, 131
Orders, 15, 22, 23
Orthoptera, 23, 130
Ovipositing, 84

Pale Evening dun, 30, 32, 41, 44, 63, 89, 94, 95
Pale Evening nymph, 139, 147
Pale Evening spinner, 29, 30, 43, 48, 64, 95
Pale Watery dun, 30, 31, 32, 37, 40, 46, 56, 90, 92, 94, 95, 103
Pale Watery nymph, 137, 146
Pale Watery spinner, 30, 43, 48, 57, 90
Palps, 19-20, 110
Paraleptophlebia, 135
Paraleptophlebia cincta, 31, 49, 72, 99, 140, 146
Paraleptophlebia submarginata, 30, 49, 71, 98, 138, 146
Perla bipunctata, 121, 122, 144, 147
Perlidae, 120
Perlodes microcephala, 121, 122, 143, 144, 147
Penultimate segments, 92
Pheasant Tail spinner, 89
Phryganea grandis, 111, 112
Phryganea striata, 111, 112
Phyllopertha horticola, 129
Plecoptera, 15, 21, 22, 120, 132
Pleskot, 142
Polluted, 108
Pond Olive dun, 30, 32, 41, 45, 64, 95, 133
Pond Olive nymph, 141, 147
Pond Olive spinner, 30, 43, 48, 65
Poplar Saw-flies, 130
Posterior, 26
Potamophylax latipennis, 111, 112
Procloëon, 135

Procloëon pseudorufulum, 30, 32, 49, 63, 94, 139, 147
Pronotum, 122.
Prontonemura meyeri, 121, 124, 145, 147
Psychomia pusilla, 111, 119
Pupal case, 18
Pupa, 107, 108
Pupae, 19
Pupate, 18, 20, 127
Purple dun, 31, 38, 47, 72, 99
Purple nymph, 140, 146
Purple spinner, 31, 41, 48, 72, 99

Red spinner, 30, 55
Reed Smuts, 18, 105, 107
Rhithrogena, 135
Rhithrogena haarupi, 30, 49, 66, 96, 139, 146
Rhithrogena semicolorata, 30, 49, 67, 81, 97, 146
Rhyacophila dorsalis, 111, 116
Ringing, 50
River Avon, 139
River Clyde, 112
River Itchen, 86, 97, 100, 128, 141
River Test, 97, 112, 121, 141
River Thames, 85
River Usk, 85, 94, 96, 100, 102, 106, 112, 123, 139, 144
River Wylye, 89, 139
Ronald's, A., 116, 130, 131
Roof-shaped wings, 15-19

Sailor Beetle, 129
Sand-fly, 111, 116
Saw-fly, 130
Sedge-flies, 22, 109, 127
Segments, 50
Sepia dun, 31, 38, 46, 75, 100
Sepia nymph, 140, 146
Sepia spinner, 31, 41, 48, 75
Sericostoma personatum, 111, 114
Sherry spinner, 28, 30, 41, 48, 61, 91, 99

INDEX

Shrimps, 22, 24, 126
Sialis spp. 127
Silo nigricornis, 111, 115
Silt Crawlers, 134, 143, 146
Silvered hooks, 104
Silverhorns, 109, 111, 118
Silver Sedge, 111, 113, 117
Size, 27, 28
Size Key (Ephemeroptera), 36
Small Brown, 121, 125
Small Brown nymph, 145, 147
Small Dark Olive dun, 30, 32, 37, 40, 46, 55, 57, 88, 90, 92, 93, 103, 132
Small Dark Olive nymph, 137, 146
Small Dark Olive spinner, 30, 43, 47, 56, 89
Small Red Sedge, 111, 119
Small Red spinner, 29, 30, 56, 89
Small Spurwing, 18, 29, 30, 32, 40, 44, 46, 61, 90, 92, 94, 95
Small Spurwing nymph, 138, 147
Small Spurwing spinner, 62, 94
Small Yellow Sally, 121, 125
Small Yellow Sally nymph, 145, 147
Small Yellow Sedge, 111, 119
Smuts, 24
Snails, 24
Soldier Beetle, 129
Solomon's Seal Saw-flies, 130
Species, 23
Spent, 29
Spent Gnat, 30, 41, 84
Spent Male, 18
Spiders, 22, 126, 129
Spinners, 17, 29, 36
St. Mark's Fly, 105
Stenophlax stellatus, 111
Sticklebacks, 24
Stone Clingers, 134, 135, 139, 141, 142, 146
Stoneflies, 21, 22, 120, 127
Stonefly creepers, 143–146
Stonefly nymphs, 121, 132, 143, 151

Striate marking, 112, 113
Sub-imago, 16
Summer Olive, 30
Surface film, 16–19
Swarms, 28, 29

Tadpoles, 24
Taeniopteryx nebulosa, 121, 124, 145, 147
Tails, 25
Tarsus, 26
Temperature, 27, 29, 97, 141
Terrestrial insects, 107, 126
Thorax, 17, 19, 25, 109, 120
Tibia, 26
Tinodes waeneri, 111, 119
Tipula spp., 107
Toothed antennae, 113
Tracheal gills, 133
Transposition, 17
Trichoptera, 15, 19, 22, 109, 119, 132
Trouts diet, 24
Turkey Brown dun, 30, 38, 46, 71, 83, 98
Turkey Brown nymph, 138, 146
Turkey Brown spinner, 30, 41, 48, 71, 99
Two Lakes, 93

Upwinged Flies, 15, 22, 24, 35

Vegetable detritus, 116
Veniards, 104
Venter, 25, 50

Wade, Dr. Michael, 28, 102, 106
Wasps, 23, 24, 130
Water Louse, 22, 131
Water Shrimp, 131
Web of silk, 115
Welshman's Button, 109, 111, 114
White Midge, 36
Willow-fly, 21, 121, 124, 125
Willow-fly nymph, 145, 147

Wing cases, 133
Wings, 25

Yellow Evening dun, 30, 38, 47, 59, 69, 90
Yellow Evening nymph, 135, 137, 146
Yellow Evening spinner, 30, 41, 48, 59
Yellow Halo, 104
Yellow Hawk, 30
Yellow May dun, 30, 38, 46, 68, 59, 90, 98,
Yellow May nymph, 139, 146
Yellow May spinner, 30, 42, 48, 69
Yellow Sally, 121, 123
Yellow Sally nymph, 144, 147
Yellow Upright spinner, 30, 42, 48, 68, 97, 98